Writing Travel

Iberian and Latin American Studies: The Arts, Literature and Identity

---

Volume 10

Edited by Professor Francis Lough
Department of Hispanic Studies, University of Birmingham

PETER LANG
Oxford · Bern · Berlin · Bruxelles · New York · Wien

Guadalupe Gerardi

Writing Travel

The Work of Roberto Bolaño and Juan José Saer

PETER LANG
Oxford · Bern · Berlin · Bruxelles · New York · Wien

Bibliographic information published by Die Deutsche Nationalbibliothek.
Die Deutsche Nationalbibliothek lists this publication in the Deutsche Nationalbibliografie; detailed bibliographic data is available on the Internet at http://dnb.d-nb.de.

A catalogue record for this book is available from the British Library.

Library of Congress Cataloging-in-Publication Data
A CIP catalog record for this book has been applied for at the Library of Congress.

ISSN 1662-1794
ISBN 978-3-0343-2215-7 (print) • ISBN 978-1-78874-216-0 (ePDF)
ISBN 978-1-78874-217-7 (ePub) • ISBN 978-1-78874-218-4 (mobi)

Hochfeldstrasse 32, CH-3012 Bern, Switzerland
info@peterlang.com, www.peterlang.com, www.peterlang.net

This publication has been peer reviewed.

A mi familia, compañeros de viaje

Contents

Acknowledgements

In completing this book, which originated when I was a PhD student at University College London, I have been very fortunate in receiving help and advice, so generously given, from a number of people. I am especially grateful to my supervisor and friend, Claire Lindsay, for her dedicated guidance, priceless support and encouragement. I would like to thank Niall Binns, Nick Caistor, Jason Wilson, Enrique Mareque, Anna Wilson, María del Pilar Blanco, Philip Swanson and especially my father, all of whom offered thought-provoking feedback, recommending material and discussing the ideas in this book. I am grateful to the anonymous reviewer from Peter Lang and to Francis Lough, who offered valuable suggestions and helped to shape the final version of this book. I must also express my appreciation of the support of the Arts and Humanities Reserch Council, whose award of a three-year doctoral studentship made the completion of the thesis possible in the first place. Any errors that are found herein are mine alone.

Introduction

In a speech delivered by Roberto Bolaño in Vienna in 2000, he stated that, for writers, the only existing motherland was their own library. In his view, he added, the commonly heard nostalgia for the motherland always sounded like a lie.[1] Bolaño then went on to question and debunk links between exile, nation and literature. As in this speech, in his fictional work Bolaño also engages with 'motherlands', and his approach to ways of conceiving national territories is at times transgressive. But despite the directness of Bolaño's remarks in his speech, and the innovative ways in which he treats the nation in his fiction, his literature remains entangled in long-standing cultural dilemmas and controversies around notions of the nation and literature in Latin America. A similar note can qualify Juan José Saer's work, which takes a specific area of Argentina as its main significant setting, while questioning the legitimacy of notions of literature in relation to the nation. For instance, in 'Literatura y crisis argentina' [Argentine Literature and Crisis], Saer affirms that great literature overcomes classificatory systems, such as that of national literature, and reduces local circumstances to just one of many literary elements. In his view, writers even create their literature from a personal stand 'una y otra vez sin tener en cuenta ni la expectativa de su audiencia ni los cambios de situación' [time and again overlooking the expectations of readers and changes of location].[2] In 'El escritor argentino en su tradición' [The Argentine Writer in His Tradition], Saer tones down the importance of the individual in forging their own literature but highlights that literary traditions are constantly changing and that loose

1 Roberto Bolaño, 'Literatura y exilio', in *Entre paréntesis* (Barcelona: Anagrama, 2004), 40–46. For more on this see Bolaño's 'Exilios', and 'Final: "*Estrella distante*" (Entrevista de Mónica Maristain)', both in *Entre paréntesis*, 49–59 and 329–344, respectively.

2 In Juan José Saer, 'Literatura y crisis argentina', in *El concepto de ficción* (Buenos Aires: Seix Barral, 2010), 94–120, 119.

social and political commonalities between cultures, such as the experience of violence and uncertainty, can create a common ground for a tradition to be used and seized by another.[3]

Notwithstanding these authors' common aversion to links between the national and their work, their territories of birth and the journeys laid across them are fundamental tropes in their literature. Moreover, both Bolaño's and Saer's oeuvres offer powerful re-readings of the narratives of travel that have influenced the territorial and cultural contours of those nations. The fiction of these two authors seems to suggest that travel and space (even national) are key in their writing, but that those features are not directly conducive to the consideration of their literature as national. Confusing, false and unrealistic are just some of the labels Bolaño and Saer hang on national classifications of literature. To be sure, their discussions seek not only to be controversial, but to broaden and reconfigure conceptions of literature as influenced by the nation, reformulating the relationship between the two, which has bound Latin American literature since the nineteenth century.

Bolaño's and Saer's reflections on the category of literature as related to the nation provide an acute example of the complexities involved in the analysis of travel and space in their work. It is in light of these debates that my book, *Writing Travel*, is located within the intersecting spatial and cultural composites that form both Bolaño's and Saer's work. Indeed, travel and space are not only a thematic part of their literature, but tropes that raise broader cultural questions. This study pursues ways of unravelling those questions. In *Writing Travel* therefore, I aim to shed light on the implications at stake in the ways in which these two major Latin American

3 In Juan José Saer, 'Literatura y crisis argentina', in *El concepto de ficción* (Buenos Aires: Seix Barral, 2010), 94–120 and subsequently 'El escritor argentino en su tradición', in Saer's *Trabajos* (Buenos Aires: Seix Barral, 2006), 61–67. In the latter article Saer responds to Jorge Luis Borges's much debated essay 'El escritor argentino y la tradición' claiming that Borges's thesis on literary traditions is correct yet incomplete. According to Saer, Borges's essay does not bear in mind the transformations that the Argentine writer effects, using local elements on the Western traditions they appropriate. Borges's 'El escritor argentino y la tradición', in *Discusión* in *Obras completas* (Buenos Aires: Emecé, 1969), 151–162.

authors, Roberto Bolaño and Juan José Saer, engage with travel and space in their literary work. The title of my study, while directly announcing its subject, aims to underscore the significance of the tropes it contains. For, in effect, in the work of Bolaño and Saer, travel and space are key structures of representation within which cultural traditions are interrogated, reassessed and reformulated. That is why the first part of the title inverts the conventional term 'travel writing', signalling this book's concern not only with how Bolaño and Saer write fictional journeys, but also how textual and physical displacement are used in these authors' novels as categories through which to dislocate traditional views of culture. My title therefore relates to travel writing but it inverts that term to reflect on the remodelling of that textual practice in Bolaño's and Saer's oeuvres.

Travel writing is commonly considered a hybrid genre or, in many instances, a category of texts that, as Jan Borm aptly notes, can include *travel literature*.[4] But the order of terms that designate that genre presupposes a referential pact (because we travel there is writing) presenting the journey as an indispensable element for writing. *Writing Travel* instead, as a reversal of those notions, does not necessarily assume a referential inscription for the journey, at the same time as it draws attention to the relevance of its compositional and close-to-literature quality (there is writing). A main aim of this book therefore emerges from the fictional quality of the corpus and is directed at scrutinizing Bolaño's and Saer's reflections on the nature of literature itself, conceived in many instances as coextensive to the journey.

While I assert the importance of the literary in my title, it is fundamental to understand the connections and disconnections my study establishes with the genre of travel writing. The immense significance and magnitude of travel writing has been identified by several scholars who have not only studied it as a textual practice that has brought the faraway into written and domestic terms, but also, and fundamentally, as a mode of colonialist discourse which has reinforced Western values and rules. For instance, Edward Said's landmark study *Orientalism* (1978)

4 Jan Borm, 'Defining Travel: On the Travel Book, Travel Writing and Terminology', in *Perspectives on Travel Writing*, eds Glenn Hopper and Tim Youngs (Aldershot: Ashgate, 2004), 13–26.

argues that travel narratives formed part of an ample body of work which participated in forming imperial structures of cultural domination and incorporation of 'what is a manifestly different (or alternative and novel) world [...] in an uneven exchange with various kinds of power'.[5] That argument has been supported by Mary Louise Pratt's *Imperial Eyes* (1992), where she argues that 'travel and exploration writing *produced* "the rest of the world" for European readership' whilst its 'signifying practices' encoded and legitimated 'the aspirations of economic expansion and empire'.[6] In consonance with the aims of my own work, Pratt's also highlights how writers 'on the receiving end of European intervention' reproduced and at times revised and transcended those models of travel writing.[7] The analysis of the connections between imperial and colonial enterprises and travel writing is likewise the subject of both David Spurr's *The Rhetoric of Empire* (1993) and Dennis Porter's *Haunted Journeys* (1991). Spurr's focus is on 'historical instances' alongside 'rhetorical functions' that formed the discourses of colonialism.[8] In a relevant manner, Porter's study alerts us to the fact that some of those writings of travel, at best, had the ability to 'overcome cultural distance through a protracted act of understanding', notwithstanding many other instances in which they contributed to the 'expression of European conceit or racist intolerance'.[9] This observation is particulary relevant to my work in that *Writing Travel* specifically engages with the reception and reformulation of the ideas of travel as a tool for reasserting Western values in the work of Bolaño and Saer.

In the Latin American context, Claire Lindsay observes that accounts of travel 'have had tremendous impact in and beyond the region itself,

5 Edward Said, *Orientalism* (London: Penguin Classics, 2003), 12.

6 Mary Louise Pratt, *Imperial Eyes: Travel Writing and Transculturation* (New York: Routledge, 2008), 4.

7 Pratt, *Imperial Eyes*, 4.

8 David Spurr, *The Rhetoric of Empire: Colonial Discourse in Journalism, Travel Writing, and Imperial Administration* (Durham, NC and London: Duke University Press, 1993).

9 Dennis Porter, *Haunted Journeys: Desire and Transgression in European Travel Writing* (Princeton, NJ and Oxford: Princeton University Press, 1991), 3.

both of a positive and negative kind'.[10] As Lindsay points out, the writings on the travels of Alexander von Humboldt and Captain Richard Burton, for example, not only 'have established and perpetuated a range of enduring myths about the continent' but also in some cases had 'direct or indirect military application'.[11] In this, Lindsay echoes Pratt's exposition that for the Latin American colonies, codifications of travel, such as those of Humboldt, fuelled the legitimation of white territorial expansion and economic domination against the Indian Other. As Pratt puts it, 'when South America's lettered elites reflected on emergent American society in the 1820s, 1830s, and 1840s' they were 'faced with the challenges of decolonizing their cultures, subjugating minorities [...] [and] projecting their hegemony into the future', for that, 'they turned with remarkable consistency to the Utopian Americanist esthetic codified by Humboldt, who had found it, in part, in them'.[12]

In a similar vein to Pratt, other critics such as Beatriz Sarlo, Jens Andermann, Beatriz Colombi and Adolfo Prieto have observed a relationship between travel writing and the creole, lettered elites' codification of national symbols and culture.[13] For instance, Adolfo Prieto observes a relationship between the writing of English travellers to Argentina in the nineteenth century and the emergence of a foundational Argentine literature. In his book *Los viajeros ingleses y la emergencia de la literatura argentina 1820–1850* (1996), Prieto demonstrates that emblematic Argentine texts appropriated symbols, images and connotations (such as the description of Patagonia as a desert and its comparisons with the sea) that were initially used by English travellers in their writings.[14] To prove his thesis, he

10 Claire Lindsay, *Contemporary Travel Writing of Latin America* (London: Routledge, 2010), 4.

11 Lindsay, *Contemporary Travel Writing*, 4.

12 Pratt, *Imperial Eyes*, 184.

13 Beatriz Sarlo, 'En el origen de la cultura argentina: Europa y el desierto', in *Escritos sobre literatura argentina* (Buenos Aires: Siglo Veintiuno Editores, 2007), 25–29; Jens Andermann, *Mapas de poder: una arqueología literaria del espacio argentino* (Rosario: Beatriz Viterbo, 2000).

14 Adolfo Prieto, *Los viajeros ingleses y la emergencia de la literatura argentina 1820–1850* (Buenos Aires: Sudamericana, 1996). As part of his examination of the texts written

analyses the influence of the texts of fourteen English travellers from the second and third decade of the nineteenth century (who went to Argentina either to look for gold mines or to explore the country both geographically and politically) to show the influence such writing had on that of some of the foremost Argentine political and literary figures such as Domingo F. Sarmiento, Juan B. Alberdi and Esteban Echeverría. Beatriz Colombi also observes a relationship between displacements and the construction of the modern imaginary at the end of the nineteenth and the beginning of the twentieth centuries in the context of Latin American literature. Colombi looks into fictional and non-fictional works of writers such as José Martí, Paul Groussac, Alfonso Reyes, and Horacio Quiroga, defining travel as either a powerful imaginary which feeds from fiction or, in a similar vein to that of Pratt's, as a 'práctica ligada a la ciencia y a la expansion territorial de Occidente' [practice related to science and to Western territorial expansion].[15]

My emphasis and intentions differ from those studies reviewed above, since my work shifts the focus to contemporary writings of fictional travel and space in the work of Bolaño and Saer. However, the analyses of those scholars outlined in the preceding paragraphs have provided instructive and illuminating models for observing how Saer's and Bolaño's narratives also find creative potential engaging with narrative precedents of travel. *Writing Travel* therefore examines the ways in which Bolaño's and Saer's texts draw on and challenge traditions of travel, their values and their ways of conceiving territories, whether national or not. In equal manner, my study seeks to understand Bolaño's and Saer's own codifications of travel and space as means of cultural self-perception and understanding.

My reformulation from travel writing to *writing travel* seeks likewise to highlight the type of engagement by Latin American writers with previous empirical travel accounts by foreigners which, as Jason Wilson notes, in complex ways, created Latin America. Wilson specifies that 'Latin America

by English travellers, Prieto studies the work of Francis Bond Head, Joseph Andrews, Edmond Temple, W. H. B Webster, Robert Fitz-Roy and Charles Darwing.

15 Beatriz Colombi, *Viaje intelectual: migraciones y desplazamientos en América Latina (1880–1915)* (Rosario: Beatriz Viterbo, 2004), 13.

is the creation of foreigners writing about the New World, from the earliest chroniclers like Bernal Díaz, to scientific travellers like Alexander von Humboldt and Charles Darwin, to later literary explorers such as D. H. Lawrence and Graham Greene'.[16] The response to those factual narratives has not generated a strong empirical tradition of internal travel writing in Latin America of the kind that dominated the outside perspective, but produced, instead, a substantial corpus of fictions within which travel writing was examined. Thus, mapping the territory from and in Latin America has been, to a great extent, a literary endeavour, and one which warrants my approach in this book: the first study of *Writing Travel* in the work of Bolaño and Saer. My book thus investigates how the novels of Saer and Bolaño rethink symbolic processes where travel and space are bound to the history and representation of their nations and Latin America.

For instance, *Los detectives salvajes* (1998) and *2666* (2004), the two of Bolaño's novels analysed here, register changes that occurred in Mexico from the 1970s, through the implementation of a neoliberal State in that country, to the devastating effects produced by the conditions of extreme capitalism and a burgeoning narco-culture in peripheral areas. Bolaño's fictions likewise engage with the changing role that literature was to play in Latin American nations since the last decades of the twentieth century. In this regard, Jean Franco aptly points out that the adoption of a neoliberal model, and the aftermaths of the military dictatorships in several countries in Latin America, not only transformed the role of the state but also the stake that literature had on definitions of the nation and community.[17] Nicola Miller agrees that, from the 1980s on, Latin American writers and intellectuals not only experienced changes in their working conditions but also in their relationship to the Nation-State.[18] As Miller points out, whether or not 'national identity could be debated in primarily cultural terms also

16 Jason Wilson, 'Travel literature', in Verity Smith, ed., *Encyclopedia of Latin American Literature* (Chicago: Fitzroy y Dearborn; Taylor and Francis, 2007), 803.

17 Jean Franco, *The Decline and Fall of the Lettered City: Latin America in the Cold War* (Cambridge, MA and London: Harvard University Press, 2002).

18 Nicola Miller, *In the Shadow of the State: Intellectuals and the Quest for National Identity in Twentieth-Century Spanish America* (London: Verso, 1999).

came under increasing scrutiny in the face of challenges to state sovereignty posed by the global media, the international debt crisis, the drugs trade, and the development of a mass politics from below, based on cross-border issues'.[19] The changes pointed out by Miller and Franco conditioned the ways in which Bolaño and Saer wrote about travel, space and nation in their literary work. Saer's novels studied here show this author's revisional bent perhaps more distinctively than Bolaño's fictions. *El entenado* (1983), *La ocasión* (1988) and *Las nubes* (1997) interrogate textual outputs that emerged as connected to historical processes in Latin America, such as conquest narratives and foundational national literature. The exploration of the relevance of displacement, migration and key spaces in the tracing of national literature and history are, as a result, woven into Saer's oeuvre and my discussion.

The most important space in the novels explored here, the desert, is read as a cultural construct to which complex meanings and themes are attached. In effect, the deserts that inspired the corpus, in Sonora and Patagonia, materially, culturally and politically constituted the frontier between 'civilization and barbarism', north and south. My study demonstrates how the work of Bolaño and Saer engages with those particular 'imaginative geographies', actively interrogating and intervening in their construction. Thus, both Bolaño's and Saer's novels set out to re-think not only the importance of travel narratives but also codifications of space in the constitution of cultural identities. Through close readings of these author's fictions and drawing from a number of theoretical frameworks from Edward Said, Caren Kaplan, Arjun Appadurai, and Homi K. Bhabha on travel and space, along with the work of cultural theorists such as Beatriz Sarlo, Fredric Jameson and Jean Franco, *Writing Travel* raises questions about the meanings that emerge from Bolaño's and Saer's engagement with preceding travelogues and their own writing of the journey and space.

Some of those questions can be sketched as follows: what readings do Bolaño's and Saer's fictions offer of writings of travel and space in Latin America? How grounded and dependant on the historical contexts are the

19 Miller, *In the Shadow of the State*, 257.

travelling experiences narrated in their own works? What types of travel are ciphered in these authors' works? How are definitions and demarcations of the nation inflected by the writing of travel and space? How do Bolaño's and Saer's fictions speak to the intersections between the writing of travel and space and colonial enterprises? How do these two authors construct their own writing of travel and the desert as instruments of transgression and critique? This book aims to offer answers to those questions by inquiring into the politics of representation in Bolaño's and Saer's work and the cultural and ideological implications at stake in 'writing travel'.

I have organized my discussion into five chapters, each centred on a specific novel. The works chosen engage differently with the issues raised above but also share pertinent areas of concern. Chapter 1, 'Desert Landscapes: A Travelling Dialogue with the Nation in Juan José Saer's *Las nubes*', illustrates how symbols that have represented national culture, such as the Patagonian Desert in Argentina, are (re)formulated in Saer's work to destabilize the idea of the national in its relation to culture and literature. My analysis also highlights how the novel emphasizes the fundamental role of heterogeneity and arbitrariness in cultural formations. Taking my cue from Arjun Appadurai and Homi Bhabha, I interpret the desert in *Las nubes* as a spatial paradigm of flexible limits which drives us to re-think the blurred contours of national cultures. I posit that the exceptionality of Saer's novel lies in the way that it dismantles foundational images and premises on which Argentine literature has been defined. Chapter 2, 'Stray Journeys: Travel in Roberto Bolaño's *Los detectives salvajes*', explores the poetics and politics involved in the writing of travel. The main questions addressed are why contemporary narratives of travel in Latin America have been traditionally interpreted as 'allegories of the national', and what the consequences of sustaining that view are. While considering those questions, this chapter examines the congruencies between writing and the politics of displacement, as well as the formal, thematic and cultural operations at stake in the crafting of the journey in *Los detectives salvajes.* Chapter 3, 'Juan José Saer's *El entenado* and the Failures of the Conquest', considers how Saer frames his 1983 novel in a dialogue with colonial discourses of travel. At stake in that text's journey are both the construction of the traveller's own identity and his dialogue with the discursive practices that resulted from the Spanish Conquest in Latin America. As the title of the chapter indicates, the narration of *El entenado* is one of failure, not only of the civilizing enterprise of the Conquest, but ultimately also of cultural understanding. This chapter analyses *El entenado*'s tension with historical discourses, alongside its take on the Spanish Conquest and some relevant concepts in this respect such as that of discovery and possession.

Bolaño's rendering of the desert in *2666* and the implications on the nature of literature in relation to that space is the focus of Chapter 4, 'Hidden in a Desert Cave: Reading and Disclosure in Roberto Bolaño's *2666*'. Therein, I examine the reading protocols created by Bolaño's masterpiece, whereby to apprehend the text is to disclose multiple meanings. I see the desert in that novel as a locus of complex, social, cultural, temporal and spatial significations, for it registers issues related to border crossings, extreme violence, femicides and poverty while simultaneously epitomizing the narrative impulse of the text in which significant meanings appear hidden. I develop an account of the complexities of understanding this contested space, a site that registers both artistic intentions and quasi-dystopian circumstances. Finally, Chapter 5, 'The Event of Literature: A Reading of the Desert and Migration in Juan José Saer's *La ocasión*', addresses how Saer engages with the complexities of migrancy while inquiring into the effects that these can have on identity. Distancing effects are put to use in this novel, both thematically and formally, to reflect on the experience of displacement. As with my analysis in Chapter 4, a fundamental issue examined here is how Saer's novel invokes and invites a particularly readerly experience, built around the conception of literature as an event.

CHAPTER 1

Desert Landscapes: A Travelling Dialogue with the Nation in Juan José Saer's *Las nubes*

In his essay, 'Literatura y crisis argentina', written in 1982, Saer discusses the difficulties that arise from defining literature in relation to the term 'nation'. As he explains, 'in art, generalizations are of no use', and encapsulating a series of works under the term 'national' creates the risk of 'transforming a purely informative and classificatory concept of "national literature" into a normative category and model'.[1] Saer's reflections are central to his aesthetics, not only because they act as a warning against 'national' readings, and in so doing intend to cancel out prescriptive views, but mainly because these provide an informative lens through which to examine his 1997 novel, *Las nubes*.[2] This text narrates the eventful story of two doctors trying to set up a psychiatric institution in the countryside in Argentina. It does so while foregrounding the importance of what the protagonist characterizes as the main event of the text: a journey through the Argentine 'desert' or Pampas. The terminus of this trip, undertaken by the main character and a caravan of madmen, prostitutes, soldiers and Indians, is echoed in the novel by an interrogation of national destinies. Thus, the journey's fate stresses a relevant question lodged at both its very core and at that of the nation: what is the role of arbitrariness

1 The original in Spanish reads: 'como en arte las generalizaciones no son de ninguna utilidad, definir la literatura en relación con el término 'nación' o con alguno de sus derivados no corresponde a ninguna realidad y en cambio presenta el inconveniente de crear toda una serie de confusiones'. Juan José Saer, 'Literatura y crisis argentina', in *El concepto de ficción* (Buenos Aires: Seix Barral, 2010), 94–120 (101). Throughout this book I have relied on published translations where indicated in the notes; otherwise, translations are my own.

2 Juan José Saer, *Las nubes* (Buenos Aires: Seix Barral, 1997). All translations into English of this novel are my own.

in reaching a destination/destiny? In this respect, *Las nubes* seems to posit a response that dismantles notions of the national as a reflection of cultural and literary constructions, highlighting the significance of randomness both in the formation of the nation and in journeying.

The relevance of randomness and travel becomes apparent from the opening of *Las nubes*. The novel starts with a brief introductory account that explains how a manuscript from the 1800s, found by chance in Argentina (and presented as the story to follow), was transported to Paris and reached the hands of Pichón Garay, a recurrent and well-known character in Saer's work. This introduction in *Las nubes* is formally differentiated from the rest of the text (it is written in italics) and it gives a description of the external conditions in which the reading of the subsequent manuscript is going to occur (location, weather, characteristics of the room and even what this character is eating). *Las nubes* is thus inscribed in a long tradition of narratives which explore the possibilities of presenting how texts have been (fictionally) encountered; that is, it constitutes a version of what we might call 'travelling' manuscripts. These texts explain their own journeying and, in so doing, they present the circumstances by which they have reached the hands of a fictional reader; more precisely, how they became readable for both the fictional, and ultimately, the non-fictional reader. Thus, the novel indicates at the outset that journeying and readability are related, and, in turn, that the manuscript's trajectory to readability is across borders.

It is symptomatic that Saer's own essay on *Las nubes* also reinforces the connection between travel and readability.[3] In that text Saer highlights the main role of travel in *Las nubes* alongside those of space and delirium. Further, the author explains that the journey through the desert in *Las nubes* connects his work with the literature written by travellers in Argentina from the end of the eighteenth century and through the nineteenth century, the appearance of which was both highly influential for founding fictions and simultaneous with the constitution of Argentina as a country.[4] This histori-

3 Juan José Saer, 'Dos razones', in *La narración-objeto* (Buenos Aires: Seix Barral, 1999), 155–166 (162).

4 In Juan José Saer's, 'La perspectiva exterior, Grombowicz en Argentina', in *El concepto de ficción* (Buenos Aires: Seix Barral, 2010), 17–29. Saer recalls the works of

cal link between the emergence of a nation and travel literature written by foreigners is key for Saer and it shows how travel writing made available readings of the nation (its geographies, landscapes, general features of the country and so on). Fundamentally, Saer's association of travel literature and the emergence of the nation to *Las nubes* alludes to how this novel is fashioned around interrogations of the connections between nation and travel. For these reasons, the first concern of this study is to show the fascinating ways in which travel is formulated and encoded in the novel. As part of this I analyse how, through *writing travel*, Saer interrogates cultural processes of nation-building within which there is both an attempt to demarcate territories and homogenize differences. My purpose is also to inquire into the diversity of Saer's literary borrowings to construct a *new literary space*, which indicate that his fiction not only stems from the inspiration of foundational texts related to national literature, as part of the existing scholarship recognizes, but that it is essentially this category of national literature that *Las nubes* is fundamentally questioning. In this respect, I offer a reading of the novel which shifts the ways in which *Las nubes* has been read to date. In essence, my reading brings to light Saer's deconstruction of the imaginings of the nation, through bringing to the fore a new *writing of travel* and the desert.

In this chapter I seek both to consider and amplify critical work on *Las nubes* that links it to Argentine foundational narratives. In this regard, Beatriz Sarlo's review of *Las nubes* needs special acknowledgement. Her work broadly traces convergences between some canonical Argentinean travel accounts and Saer's novel, but stops short of explaining how Saer transgresses these common characteristics (and this is a central point of my analysis). She starts her review by rightly observing that *Las nubes* allows the reader to study Saer's 'modo compositivo' [mode of composition][5] with

---

'travel writers' in Argentina such as Félix de Azara, Millau, Mac Cann, Woodbine, Hinchliff, Alfred Ebelot, Albert Londres and W. H. Hudson', 'esa literatura de viajeros es contemporánea a la aparición misma del país' [that literature of travellers is contemporary to the very coming into being of the country] (20).

5 Beatriz Sarlo, 'Aventuras de un médico filósofo', in *Escritos sobre literatura argentina* (Buenos Aires: Siglo XXI editores, 2007), 296–303 (297).

relation to literary traditions and, within them, Argentine literature. Sarlo positions *Las nubes*' account of the voyage within the tradition of travel writing because, as she puts it, 'hay un saber adquirido en los viajeros que visitaron el Río de la Plata en el siglo XIX que Saer utiliza en la redacción del viaje imaginario de *Las nubes*' [to write *Las nubes*' imaginary trip, Saer uses knowledge gained from the travellers that visited the River Plate in the nineteenth century].[6] Likewise, in her view, there is an affiliation between *Las nubes* and the main texts written by some emblematic Argentinean writers. For example, Sarlo sees similarities between Saer's Indians and some characters of Lucio V. Mansilla's *Una excursión a los indios Ranqueles* (1870); and between Saer's themes and some of the motifs that appear in Ricardo Güiraldes' *Don Segundo Sombra* (1926) and in Benito Lynch's *Raquela* (1918). Sarlo also notes that some of the shared tropes between the work of these writers and Saer are the short stories recounted round the campfire, alarming rumours about the Indians, the portrayal of native types, the extension of the landscape, the illusion of horizontality, and the description of natural catastrophes. Sarlo considers that Saer resorts to his reading of Domingo Faustino Sarmiento in the novel, since both include the viewpoint of a traveller who goes through familiar territory but sees it from the perspective of someone who has been outside of his own country.

In a similar vein to Sarlo, Isabel Quintana (in her work entitled 'Narrar el desierto') notes in Saer's novel the creation of an ideal space that relates to canonical historical writing. In Quintana's own words, the desert is 'un escenario abierto e incluso utópico con relación a la historia nacional' [an open and even utopian setting related to national history].[7] Likewise, Julio Premat's analysis of *Las nubes* coincides with Sarlo's and Quintana's view, in that he proposes that the novel is analogous to classical texts, such as Virgil's *Aeneid* and *The Fourth Bucolic*, in the way that they all speak to the formation of a new nation. Premat considers that Saer's literature is to a large extent a re-writing of initiatory texts since the author incorporates

6 Sarlo, 'Aventuras de un médico', 297.

7 Isabel Quintana, 'Narrar el desierto', in *Figuras de la experiencia en el fin de siglo, Cristina Peri Rossi, Ricardo Piglia, Juan José Saer, Silviano Santiago* (Rosario: Beatriz Viterbo, 2001), 127–137 (138).

into his literature 'mitos de variado valor fundacional' [myths of varied foundational value].[8]

I agree with Premat, Sarlo and Quintana that in this novel there is a preoccupation with discourses that were suitable for the formation of a new nation. The work of Virgil, which the main character of *Las nubes* relates to each of the vicissitudes of their trip, is a good example. But the dialogue that the novel establishes with foundational narratives is nuanced, more a calling into question of the premises of the national than a re-writing. For instance, the idea of arbitrariness affecting the formation of the nation, which was incompatible with goals of national consolidation prominent in foundational and programmatic literature, is central to *Las nubes*.[9] Thus, in the introductory text of Saer's novel the manuscript to follow is presented to Pichón Garay with one sentence that reads: '*Salió a buscar Troya y casi se topa con el Hades*' [He left looking for Troy and almost runs into Hades].[10] The first part of this phrase presents the manuscript as the story of an epic (national) journey similar to that of *The Aeneid* in which the main character, Aeneas, ventures to Troy, but the second part of the sentence refers to the role of randomness in this search, and how the journey could well have ended in the underworld. This idea is reintroduced in the middle of the novel when the main character speaks about his preference for *The Fourth Eclogue* and states that 'la nueva edad de oro no será un premio o una conquista, sino un don injustificado del destino y advendrá, no porque los hombres se la hayan ganado, sino porque las Parcas, un día cualquiera, por puro capricho, dirán que sí' [the new Golden Age will not be a prize or a conquest, but an unjustified gift from destiny and it will come, not because men have won it, but because the Fates, on a random day, on a whim, will

8 Julio Premat, 'Mitos (*El limonero real, La ocasión, La pesquisa, Las nubes*)', in *La dicha de Saturno* (Rosario: Beatriz Viterbo, 2002), 330–344 (343).

9 For more on this see Tulio Halperín Donghi's 'Sarmiento y el historicismo romántico' (Alicante: Biblioteca virtual Miguel de Cervantes, 2011) <http//www.cervantesvirtual.com/nd/ark:/59851/bmcws9b0> [accessed 12 December 2016], and Doris Sommer's *Foundational Fictions: The National Romances of Latin America* (Berkeley and Los Angeles: University of California Press, 1991).

10 Saer, *Las nubes*, 12 (original emphasis).

say yes].[11] The Parcae (or Fates) decide, capriciously, that a Golden Age of a nation is to come, and so it does. Thus, this fragment of the novel reinforces the idea that arbitrariness has much to do with national destinies. At the end of the novel, when describing the culmination of their arduous journey, this thread is picked up again. On reaching their destination, the main character comments on how their safe arrival is really just down to chance, stating: 'las Parcas, por esta vez, dijeron que sí' [the Fates, for this time, said yes].[12] It is apparent, then, that the association with *The Fourth Eclogue* in *Las nubes* does not just celebrate nor is it part of a re-writing of a foundational narrative, but it does highlight the role of arbitrariness in both the journey and the nation's destiny.

My own reading of *Las nubes* therefore supplements the important views detailed above of how the novel has a relationship with national and foundational literatures. In fact, the purpose of what follows is to argue how Saer's text engages with foundational narratives; however, not in order to continue with that tradition, or to present a utopian side to it, but, as I mentioned above, to dismantle the idea of literature defined in terms of the national.

A dialogue with history

A central feature that attests to Saer's reading and deconstruction of the category of the national in literature, is the fact that the novel takes place at the beginning of the 1800s, precisely when, as noted by José Carlos Chiaramonte, Tulio Halperín Donghi and Benedict Anderson 'early conceptions of [...] nation-ness' were being developed in Latin American countries.[13] Indeed, the events recounted in the second text of *Las nubes*

11 Saer, *Las nubes*, 115.

12 Saer, *Las nubes*, 206.

13 Benedict Anderson, *Imagined Communities: Reflections on the Origin and Spread of Nationalism*, revised edition (London and New York: Verso, 1991), 50. For more on

reach back to 1804, thirty years before the time of the fictional writing of the manuscript. These events take place in a pre-revolutionary and post-revolutionary viceroyalty (the text is set in the thick of the historical and social upheaval of Argentina's process of independence), but the fictional writing is done in Europe when Argentina is emerging as an independent nation, after the Revolution of 1810 and its official declaration of independence in 1816.

The second text of *Las nubes*, the manuscript, begins with the voice of Real, the novel's main character. He is an Argentine doctor, young and and well travelled. He gives an account of the eventualities surrounding the establishment of an asylum for the mentally ill, *Las Acacias*, in the Argentine Pampas. With his associate, Dr Weiss, he aims to found the institution and to create a therapeutic environment where they will employ new and alternative methods to cure mental illnesses. The methods he describes are influenced by the psychiatric movement that was then taking place in Paris (in a post-revolutionary France), and are opposed to those of enclosure and punishment. As we learn, both characters were stimulated by these new ideas while they lived in France before returning to Argentina to begin this novel and revolutionary enterprise. In setting up the mental institution, Dr Real has to collect five patients from a post located 100 leagues away and then take them to the asylum, accompanied by a caravan of varied characters. The protagonist-narrator recounts the journey in the tradition of travel as travail, by describing their trip through the desert as 'cien leguas de vicisitudes' [a hundred leagues of vicissitudes].[14] Due to unforeseen circumstances, the journey – that was supposed to take fifteen days – lasts six weeks. Saer contextualizes this voyage with some facts taken from Argentine history, a time when the relationship between the Spanish Crown and the colonies was going through significant difficulties.

---

this see Tulio Halperín Donghi's *Proyecto y construcción de una nación, 1846–1880* (Caracas: Biblioteca Ayacucho, 1980) and Carlos Chiaramonte's *Nación y estado en Iberoamérica: el lenguaje político en tiempos de las independencias* (Buenos Aires: Sudamericana, 2004) and *Ciudades, provincias, estados: orígenes de la nación argentina 1800–1846* (Buenos Aires: Emecé, 2007).

14 Saer, *Las nubes*, 19.

Las nubes depicts these historical tensions, particularly when describing the creation of the new mental hospital in the context of a crumbling colony: 'la empalizada con que la Metrópoli pretendía aislarlas [a las colonias] estaba agujereada por todos lados, de modo que era de lo más fácil colarse por los huecos, y hasta los que habían sido nombrados por Madrid para impedirlo se beneficiaban con la situación' [The palisade used by the Metropolis to isolate the colonies had cracks everywhere, so it was very easy to traverse those cracks, and even the people from Madrid who were hired to prevent this were benefitting from the situation].[15] The two doctors, both professed revolutionaries, experience problems from both sides of the divide: not only from groups that objected to the revolution but also the Creole post-revolutionary government after 1810. Despite these political controversies, the asylum works for a few years until Dr Weiss is found out to have had an affair with the wife of a jealous military official, who then aims to sabotage their psychiatric enterprise. The mental asylum is then attacked and destroyed, and the doctors end up fleeing to Europe.

The specific features of Saer's dialogue with founding fictions can be gathered not only from its setting in this historical period, but also, as I explain in what follows, from Saer's systematic upsetting of the cultural and geographical premises, such as the desert, that were an essential part of Argentine foundational narratives. In a study on the genesis of Argentine literature, Beatriz Sarlo confirms the complex relationship between that space, native populations and culture.[16] Sarlo explains that until the beginning of the twentieth century, intellectuals and the political establishment in Argentina equated 'the others' (Indians and Gauchos), and their territory, to the trope of the desert. Indeed, as Sarlo points out:

> La palabra desierto, más allá de una denominación geográfica o sociopolítica, tiene una particular consecuencia: implica un despojamiento de cultura respecto del espacio y los hombres a los que se refiere. Donde hay desierto, no hay cultura; el Otro que lo habita es visto precisamente como Otro absoluto, hundido en una diferencia intransitable.

15 Saer, *Las nubes*, 22.

16 Beatriz Sarlo, 'En el origen de la cultura argentina: Europa y el desierto', in *Escritos sobre literatura argentina* (Buenos Aires: Siglo Veintiuno Editores, 2007), 25–29 (28).

> [The word desert, apart from its geographical or socio-political denomination, has a particular consequence: it entails a divestment of culture with regard to the space and the men to whom it refers. Where there is desert, there is no culture; the Other that inhabits it is precisely seen as an absolute Other, sunken in an impassable difference.][17]

The Argentinean Patagonia and surrounding lands inhabited by the Indians were defined as 'the desert' by the political and intellectual establishment. Conceptualized as 'empty spaces', these regions were susceptible to being incorporated into the new nation without recognition of the territorial rights of the Indians who occupied them. This ideological enterprise reached its climax in 1879 with General Roca's military campaign, 'La conquista del desierto' and the final consolidation of Argentina as a Nation-State. The military 'conquest of the desert' resulted in the extermination of a great part of the Indian population (approximately between 55,000 and 100,000 Indians) and the segregation of the survivors. Since their lands were defined as 'deserted' and consequently absent of culture, civilization and habitation, they were therefore aligned with, or were used as, what Sarlo calls a 'nada a partir de lo que fundar una literatura y una cultura' [nothingness from where to found a literature and a culture].[18] Jens Andermann reinforces Sarlo's idea of foundational literature in Argentina as built on the space of the desert when he writes that: 'En un primer instante se busca inscribir una letra portadora de un discurso civilizador y universalista en un espacio concebido como desértico y vacío' [In this first instance what is pursued is to inscribe a writing that is the carrier of a civilizing and universalist discourse – in a space conceived as deserted and empty].[19] Thus, both scholars sustain the view that texts which emerge at the beginnings of Argentinean literature, such as Esteban Echeverría's *La cautiva* (1837) and *El matadero* (1838) or Sarmiento's *Facundo, civilización y barbarie* (1845), transform the desert (and by extension the Indian territory) into a crucial cultural and political space.

17 Sarlo, 'En el origen', 25.

18 Sarlo, 'En el origen', 26.

19 Jens Andermann, *Mapas de poder: una arqueología literaria del espacio argentino* (Rosario: Beatriz Viterbo, 2000), 19.

Such considerations serve to underline the fact that the texts of travellers in Argentina at the beginning of the nineteenth century, together with Argentine foundational narratives, establish a literary lineage in which travel and desert are at their core. What emerges at first glance when reading *Las nubes* is, therefore, that – time apart – its plot (and its caravan travelling through the desert) aligns it shoulder to shoulder with the texts written at the emergence of Argentine literature. In what follows I show that if Saer makes use of the readings of these national texts for the writing of *Las nubes*, he does so in order to grapple with, and not to reaffirm, that tradition. In effect, the novel repeatedly questions images and representations that were an integral part of the national literature.

We find a clear instance of Saer's interrogation of foundational texts in the representations of the desert and the Indians in *Las nubes*. It is important to bear in mind that, in the 1800s, not only were the desert and the Indians seen as coterminous, as Sarlo noted in the above quotation, but also the frontier character attributed to the land inhabited by the Indians. This land, more specifically termed the 'llanura' or 'desierto', was a physical boundary that separated Creole and European settlements from Indian territory, and it was used in the emblematic text by Sarmiento, *Facundo*, to set out the distinction between civilization and barbarism. Sarmiento based his argument on ideas of geographical determinism: for him the vast plains of Argentina were inhabited by the uncivilized and uncouth, as opposed to the cities, which he thought were populated by refined urbanites and where progress was possible. In this regard, Graciela Montaldo aptly points out that Sarmiento 'concibe los males políticos como males del espacio, de un territorio que es pura extensión bárbara e inconmensurable' [understands political failings as the spatial failing of a territory that is purely a vast and barbarian extension].[20] Thus, the city of Buenos Aires stood for the 'civilized' and the Pampas for 'the barbarians'. This portrayal, as Ernesto Livon-Grosman has noted, informed to a great extent the intellectual establishment's preoccupation with the definition of a national identity

20 Graciela Montaldo, *Ficciones culturales y fábulas de identidad en América Latina* (Rosario: Beatriz Viterbo Editora, 1999), 47.

in Argentina.[21] In effect, the binary civilization-barbarism not only conditioned the parameters with which the land of the gauchos and Indians was to be seen, but it also determined Argentine political and social policies for centuries to come.

With regard to the above classifications, *Las nubes* acquires a burlesque tone. For example, although the convoy dreads an attack from the Indian-Cacique Josesito and his group, the representation of that character (as an ex-Catholic, with a Christian and diminutized name, resentful of civilization but an expert violinist who plays beautiful classical pieces after they slaughter the groups of people they come across) clearly mocks the characteristics that were attributed at the time to the Indians (primitive, natural beings, lacking in culture). In *Las nubes*, although Josesito's deeds take place in the desert (conceived at that time as a 'natural' space), he is not characterized as 'natural', and his actions proceed from his resentment towards culture and cultured people. Similarly, the irrational force that drives him to slaughter the desert travellers is not 'primitive' but culturally and socially constructed: 'era a causa de una especie de querella religiosa que Josesito había desertado de la civilización declarándole la guerra a los cristianos' [It was due to a kind of religious dispute that Josesito had deserted from civilization, declaring war on the Christians].[22] In the same manner, the depiction of the assimilated Indian called Sirirí (who travels in the convoy) is of a Christian, exaggeratedly strict in his morals, and almost a counter-parody of the negative image of the Indian that emerged in Argentina after 1830.[23] Accordingly, the descriptions of both Indians, Josesito and Sirirí, show the strength of the connection between culture and nature, and hence undercut Sarmiento's divisions between civilization and barbarism.

21 Ernesto Livon-Grosman, *Geografías imaginarias: el relato de viaje y la construcción del espacio patagónico* (Rosario: Beatriz Viterbo, 2003), 25.

22 Saer, *Las nubes*, 62.

23 For more on this see David Viñas, *Indios, ejército y frontera* (Buenos Aires: Santiago Arcos Editor, 2003), 70. Here Viñas explains the changes that occurred in the vision of the Indians after the independence period of 1810–1830. When the Creole verified the Indians' lack of interest in their revolutionary call, they started to characterize them as 'disastrous beasts', 'bloody tigers', and other designations that were previously used to denounce the Spaniards in America.

Likewise, the descriptions of the desert in *Las nubes* present this territory as a locus of knowledge, thereby calling into question the characteristics that the intellectual establishment of the time (e.g. Sarmiento) attributed to the 'tierra vacía': the land of ignorance and barbarism, where no knowledge was possible.[24] *Las nubes* insists that there is a knowledge obtained from living in the desert, as well as a knowledge that one must count on to live in this territory:

> El innegable saber de Osuna en todo lo relativo a la inmensa llanura [...] lo ponía en una situación ventajosa pero incómoda, que el sabio o el artista quizás puedan comprender ya que, semejante a la ciencia del desierto que practicaban Osuna u otros como él, el sabio y el artista deben tratar la mayor parte del tiempo con gente que, si bien saca provecho de su actividad, es incapaz de estimarla en forma correcta.
>
> [Osuna's undeniable knowledge in everything related to the immense plain [...] put him in an advantageous situation but nonetheless an uncomfortable one, that maybe the wise man or the artist would be able to understand since, similarly to the science of the desert that Osuna and others like him practised, the wise man and the artist, for most of their time, must deal with people who, though they benefit from their activity, are incapable of holding it in the correct esteem.][25]

As opposed to 'ignorance', which was conventionally used to characterize the inhabitants of the plains, Saer presents them as highly knowledgeable. The knowledge, in part geographical, that the character Osuna possesses is extremely necessary to survive in this terrain and it is equated to the skills of the artist and the wise man. Ultimately, what matters about these characterizations is that they efface the old categories used to define the plains and its inhabitants. From this it follows that in *Las nubes*, the desert should not be seen as a borderland between civilized life and savagery.

24 As has been pointed out by David J. Weber and Jane M. Raush in *Where Cultures Meet: Frontiers in Latin American History* (Wilmington, DE: SR Books, 1994): 'In his disdain for frontier peoples, Sarmiento typified nineteenth-century Latin American intellectuals, most of whom came from the urban elite and equated cities with progress and frontiers with ignorance and primitivism', XVIII.

25 Saer, *Las nubes*, 85.

Saer's preoccupation with the representation of the Argentine plains and how these were incorporated into foundational literature is also present in his review of Alfred Ebelot's reissue of *La Pampa* (1890).[26] It is worth considering this review here, for I believe we can extract from this piece Saer's own stance on the desert as a main geographical and cultural space. Saer notes that the vision that Ebelot offers in his texts is of someone who considered the years spent in Argentina as the most important experience of his life. Ebelot was both contemporary to and a participant in the last stages of the 'Conquista del desierto' [Conquest of the Desert]. He arrived in Argentina as a civil engineer from France hired by the Argentine government and went on many excursions into the Pampas. One outcome of Ebelot's trips to the Pampas was a series of articles 'escritos en el corazón de la llanura' [written in the heart of the plains] which were characterized by their alluring descriptive power.[27] According to Saer:

> Sus descripciones de indios, gauchos y soldados, de fortines y de oficiales perdidos en las líneas de fronteras, de los viajes en diligencia a través del desierto, de las mujeres de la soldadesca, de traficantes y politicastros, saben crear *esa atmósfera singular que, a mi juicio, sólo se encuentra en los mejores films de John Ford.*
>
> [His descriptions of Indians, gauchos and soldiers, of strongholds and of officials lost amongst the frontier boundaries, of the trips by stagecoach through the desert, of the women of the licentious soldiery, of traffickers and bad politicians, know how to create *that unique atmosphere that, in my view, is only seen again in the best films of John Ford.*][28]

I believe it is particularly telling how Saer both stresses the 'unique atmosphere' of Ebelot's descriptions of the Argentine Pampa and compares them to Ford's films. Indeed, I would argue that Saer's decision to compare Ebelot's descriptions of the Argentine desert to the evocative atmosphere of John Ford's Westerns and not to Argentine foundational literature such as Sarmiento's or Mansilla's texts, nor to the literature written by English travellers prior to the creation of Argentine foundational literature, is

26 Juan José Saer, 'Ebelot', in *El concepto de ficción* (Buenos Aires: Ariel, 1997), 71–78.
27 Saer, 'Ebelot', 75.
28 Saer, 'Ebelot', 75 (emphasis is mine).

striking enough to warrant analysis. Accordingly, in what follows I aim to show that, in order to capture that arresting 'sense of the desert' found in *Las nubes*, Saer not only incorporated and contested Argentine foundational imagery, but also drew on Western films such as *Stagecoach* (1939), *Wagon Master* (1950), *The Searchers* (1956), and *Fort Apache* (1948).

In Westerns, as in Saer's novel, the desert is portrayed as a vast natural territory, a space to master, defined and characterized by its openness. Equally, as Kim Newman notes, 'the notion of the journey is central to many western sub-genres: the wagon train movie [...] the drifting hero film, the outlaw or Indian pursuit cycle', explaining that travel 'serves to isolate the characters from society at the same time as it binds them together as a miniature social group, and to pit them against the rigours of the landscape'.[29] The primacy of the journey and a comparable process of isolation are both observable in *Las nubes*. In this way, Dr Real emphasizes the journey as the main event of his account. Likewise, Dr Real and his convoy could be seen as cut off from the villages or settlements they go through during their trip, and they are conceived as a separate social group which must face the hostilities of a journey through the desert. Moreover, some of the members of the travelling convoy in *Las nubes* clearly echo characters from certain Westerns. For instance, the film *Stagecoach*, whose plot is also based on the trip of a group through the desert, includes amongst its travellers, a prostitute, an alcoholic doctor, a gambler, a salesman and a group of soldiers who follow them on horses in order to protect them from the Indian attack headed by Geronimo. These personalities have more or less direct parallels in *Las nubes*' convoy which comprises prostitutes, soldiers, the Vasc who sells food and alcohol to the travellers, Dr Real and Josesito, the Indian. Furthermore, there are scenes in *Las nubes* which are strikingly reminiscent of several Westerns. To give just one example, the passage in *Las nubes* which describes where Dr Real and a couple of characters return to their stagecoach after rescuing one of the madmen from the Indians is remarkably similar to a scene in *Wagon Master*: catching sight of Dr Real and his companions the bugler announces their return with a trumpet fan-fare after which a party ensues with dancing around the campfire:

29 Kim Newman, *Wild West Movies, or How the West Was Found, Won, Lost, Lied About, Filmed and Forgotten* (London: Bloomsbury, 1990), 46.

> Hubo una verdadera fiesta, que duró casi hasta el amanecer: ruidosos, achispados por el alcohol que el vasco, para asombro general, distribuyó gratuitamente, cantamos y bailamos en la noche sofocante, a la luz de una gran hoguera, irrisorios y diminutos, atrapados en la triple infinitud del campo, de la noche y de las estrellas.
>
> [There was a proper party, that lasted almost until dawn: noisy, tipsy because of alcohol that the Vasc, to general amazement, distributed freely, we sang and danced in the stifling night, in the light of a great bonfire, risible and tiny, trapped in the three-fold vastness of the country, the night and the stars.][30]

Likewise, the quasi-cinematographic descriptions of certain characters in the novel, such as the silhouette of the gaucho Osuna (the path-finder throughout their trip) 'galopando paralelo al sol naciente' [galloping parallel to the rising sun],[31] which the protagonist recalls and presents as recurrent and 'primordial', are highly evocative of archetypal scenes from the Westerns.

Filmic and literary images are continually intertwined in the novel, both thematically and formally. Indeed, the creative crossings between genres affects the spatial design of the desert, presenting it to the reader as a territory that seems to be constantly repeating itself:

> Durante leguas y leguas, el desierto es en cada una de sus partes siempre idéntico a sí mismo. Únicamente la luz cambia [...] Los ruidos rítmicos del desplazamiento, en carro, en carreta, en coche de postas o caballo, repitiéndose idénticos durante largos trechos, a causa de la regularidad, cuando no la ausencia, de los accidentes del terreno, parecen repetir también al infinito el mismo instante, como si la cinta incolora del tiempo, atascada en la muesca de la rueda o quién sabe qué que la desplaza, titilara en un punto inmóvil.
>
> [For miles and miles, the desert is, in each of its parts, always identical to itself. Only the light changes [...] the rhythmical noises of the movement, on cart, on wagon, on carriage or horse, exactly repeating themselves through long stretches, due to the regularity or the absence of the accidents of the terrain, seem to repeat endlessly the same instant, as if the colourless film of time, stuck on the notch of the wheel or who knows what it is that moves it, twinkled in a motionless point.][32]

30 Saer, *Las nubes*, 185.
31 Saer, *Las nubes*, 59.
32 Saer, *Las nubes*, 153.

As a cinematic and iterating image, the desert eludes fixed demarcation. In addition, the description offered in the above quotation (of the impression produced by the trip through a desert that always seems identical to itself) invokes the form of Saer's novel, since the manuscript of *Las nubes* is written as a long take without chapters or divisions; the writing itself thus very much mimics a panorama of a journey through the desert.

By making these analogies with cinematic images, specifically from Westerns, I do not wish to imply that Saer 'Americanizes' the desert in *Las nubes*. Rather, that by the invocation of Western films, the intersections of literary and cinematic forms, and the contamination by images from different geographical deserts for effect, he is questioning the national specificity of representations of the desert. In fact, Saer is showing how these representations are influenced by other, diverse, narratives and cultures, and therefore exist outside an exclusive or narrow paradigm of nation-ness. Saer's desert in *Las nubes* not only recalls that foundational texts were informed by European travel writing, but also that diverse influences are continuously permeating into key national imageries. In this sense, as Edward Said has written in a different context, *Las nubes* suggests that cultures, 'are *always* made up of mixed, heterogeneous, and even contradictory discourses, never more themselves in a sense than when they are not just being themselves'.[33]

Flexible landscapes, heterogeneous populations

The cinematic portrayal of the desert (as explained above) also reinforces the impression of this space as one with blurred borders, typifying the spatial configurations at stake in the novel. As I argue in detail in what follows, the desert in *Las nubes* constitutes a new type of *flexible landscape*. The view I am suggesting will seem clearer if we turn briefly to Arjun

33 Edward Said, *Reflections on Exile and Other Literary and Cultural Essays* (London: Granta Books, 2001), XV (emphasis is mine).

Appadurai's reflections on cultural spaces in *Modernity at Large* (1996). Here, Appadurai identifies 'how electronic mediation and mass migration mark the world of the present' for they impart motion to cultural material (like the diskette that contains the manuscript of *Las nubes*) that 'may be seen to be moving across national boundaries'.[34] Appadurai understands the 'configuration of cultural forms [...] as possessing non Euclidean boundaries, structures or regularities',[35] and suggests new, flexible, spatial paradigms or *landscapes* of fluid and irregular contours, to think about the composition of modern subjectivities. Although Appadurai's specific spatial paradigms do not dovetail exactly with *Las nubes*' spatial practices, his reflections on the subject provide a useful paradigm with which to investigate the novel's composition of spaces as having mobile contours, and not being as fixed as national frontiers are traditionally conceived.[36]

In fact, the idea of *flexible landscapes* speaks to my analysis of spaces in *Las nubes*, not only with respect to the desert but also to the clouds, whose metaphorical centrality is highlighted in the very title of the novel. Interestingly, the only reference to the word 'nubes' [clouds] in Saer's text occurs in a section during which the caravan traverses territories of extreme heat making them anxious for rain. When the convoy finally see clouds appearing in the sky, they misinterpret them as signs of a gathering storm only to realize later that they are in fact clouds of smoke from a raging fire that is advancing towards them. By bringing the contrary to what was hoped for and expected, the clouds become another symbol of arbitrariness. Despite their rare appearances on the pages of *Las nubes*, clouds are of paramount importance in the novel not only for their arbitrary nature (and we could relate this to the novel's reflections on national destinies),

34 Arjun Appadurai, *Modernity at Large: Cultural Dimensions of Globalization* (Minneapolis: University of Minnesota Press, 1996), 4 and 46 respectively.

35 Appadurai, *Modernity*, 46.

36 In his study on *Frontiers: Territory and State Formation in the Modern World* (Cambridge: Polity Press, 1996) Malcolm Anderson aptly notes that frontiers, despite not being 'part of an immutable natural order', in some cases 'acquired a mythic significance in building nations and political identities becoming the *mythomoteur* of a whole society', 1–4.

but also because they offer another crucial example of a space or an entity whose contours are blurred. The protagonist of *Las nubes* reflects on the nature of the clouds and states that they seem to be of an 'esencia semejante a la del acontecer, que va desenvolviéndose en el tiempo' [essence similar to that of occurrences, which gradually develop in time].[37] One sees the clouds, the protagonist continues, 'con la misma familiaridad extraña de las cosas que, en el instante mismo en que suceden, se esfuman en ese lugar que nunca nadie visitó, y al que llamamos el pasado' [with the same strange familiarity of things that in the very instant when they occur, vanish into that place that no one ever visited, which we call the past].[38] The shape and 'essence' of the clouds is simultaneously demarcated and overflown. They can only be traced fleetingly for, as the quotation reads, in 'the very instant they occur, they vanish'. Moreover, the words of the protagonist effectively expose the interweaving of the nature of the clouds and that of the past. In this sense, his words suggest that both the clouds and the past are not only related but 'disseminated'. The question of the dissemination of space and time also comes to the fore in the introductory text of the novel, where Pichón Garay reflects on the characteristics of both spaces and history:

> Lo que es válido para un lugar es válido para el espacio entero, y ya sabemos que si el todo contiene a la parte, la parte a su vez contiene al todo. No lo hago con veleidades de historiador porque no tengo ninguna fe en la historia. No creo ni que pueda servir de modelo para el presente, ni que podamos recuperar de ella otra cosa que unos pocos vestigios materiales, lápidas, imágenes, objetos y papeles en los que, lo reconozco, lo que aparece escrito puede ser un poco más que materia. Lo que percibimos como verdadero del pasado, no es la historia, sino nuestro propio presente que se proyecta a sí mismo y se contempla en lo exterior.
>
> [What is valid for a place, is so for the whole space. And we know that if the whole contains the part, the part, in turn, contains the whole. I do not state this with the caprices of a historian, because I do not have any faith in history. I do not think it can be used as a model for the present, nor that we could recover from it more than a few material traces, gravestones, images, objects and papers in which, I have to admit, what appears written could be just a bit more than matter. What we perceive

37 Saer, *Las nubes*, 193.
38 Saer, *Las nubes*, 193.

> as true of the past, is not history, but our own present, projected onto itself and contemplated from the exterior.][39]

In this quotation, history appears as disassembled, embodied in heterogeneous fragments, articulated as a discourse of interruptions. History is thought of as traces, grounding its spatial dimension while undermining its chronological one, as Pichón Garay says, 'it does not constitute a model for the present', for its fragments are of the present. We could 'contemplate' the past exteriorly, as if it was a text of remains. In this sense, the above words foreground what Claudio Canaparo has seen as a particular feature of thinking 'spatially', for though the extract acknowledges that 'chronologies and historical periods exist', it shows that 'in practice all aspects and periods *conviven* (are contemporaneous) in a *locus of enunciation* that authors normally refer to as "present"'.[40] The narratives of the past and of space in *Las nubes* are seen from the present and continually erase their own boundaries, distressing the possibility of adopting a fixed structure.

In relation to the design of spaces as *flexible*, one might also add, from reading Homi Bhabha, that Saer's novel is also consistent with narratives of difference in which:

> We are confronted with the nation split within itself, articulating the heterogeneity of its population. The bared Nation *It/Self*, alienated from its eternal self-generation, becomes a liminal signifying space that is *internally* marked by the discourse of minorities, the heterogeneous histories of contending peoples, antagonistic authorities and tense locations of cultural difference.[41]

The caravan of madmen, soldiers, prostitutes, a nymphomaniac nun, Indians and so on in *Las nubes* could certainly be 'articulating the heterogeneity of its population'. The stories of each of these characters are, as the lengthy and detailed descriptions of their different types of madness testify in the

39 Saer, *Las nubes*, 14.

40 Claudio Canaparo, *Geo-epistemology: Latin America and the Location of Knowledge* (Oxford: Peter Lang, 2009), 102.

41 Homi K. Bhabha, 'Dissemination: Time, Narrative and the Margins of the Modern Nation', in *The Location of Culture* (New York: Routledge, 2008), 212.

novel, simultaneously heterogeneous and liminal. These characters epitomize what Sally Shuttleworth observed in relation to the role of fools in mid-nineteenth-century Europe: they are figures 'of ontological indeterminacy, not only crossing boundaries, but troubling and disturbing them', yet central to 'attempts to define the boundaries of social normality'.[42] The mad person challenges the limits of society by existing both within it – by their being granted some civil rights – and outside of it – their political rights are withdrawn from them and they are exempt of moral responsibility. In this sense, not only are they simultaneously outsiders and insiders, they constitute an 'absolute other' which troubles the confines of citizenry and normality and hence cannot be included in a homogeneous group. In *Las nubes* the madmen problematize notions of equality (within the 'civilized') which were a great source of controversy during and after Argentina's process of independence.[43] They also challenge the processes of homogenization which became the fundamental practices of nation-building, such as those highlighted by Eric Hobsbawm: the celebration of shared linguistic, racial and other 'common' traits and the failure to recognize differences and conflicts within national economies.[44]

Historically, when faced with the binary division between civilization (city) and barbarism (countryside/desert), the nation-building processes in Argentina have excluded or eliminated the latter in favour of the former. In this context, the question *Las nubes* seems to be positing is, what is the place for these mentally ill but apparently 'civilized' people? They are the radical and, individualized by their own madness, they embody 'difference' and therefore complicate their inclusion into or the formation of any homogeneous 'civilized' identity. Although in

42 Sally Shuttleworth, '"So Childish and So Dreadfully Un-Childlike": Cultural Constructions of Idiocy in the Mid-Nineteenth Century', in *Crossing Boundaries: Thinking Through Literature* (Sheffield: Sheffield Academic Press, 2001), 17–44 (17–18).

43 For more on this see Tulio Halperin Donghi, *Revolución y guerra: Formación de una elite dirigente en la Argentina criolla* (Buenos Aires: Siglo veintiuno, 2005).

44 Eric Hobsbawm, *Nations and Nationalism since 1780: Programme, Myth, Reality* (Cambridge: Cambridge University Press, 1990).

the novel one of the doctors states that sometimes it is difficult to differentiate those who are patients from those who are not, what is being reinforced here is that even the sane could be characterized by their own 'difference'. In any case, the very diverse types of madness in the novel allow for what Bhabha sees as 'the intervention of the Third Space of enunciation', in the sense that their individual histories 'quite properly challenge[s] our sense of the historical identity of culture as a homogenizing, unifying force, authenticated by the originary Past, kept alive in the national tradition of the People'.[45]

The emphasis on the marginal position of the mentally ill is reinforced both conceptually and physically in the novel. A clear example of the liminal space the patients occupy is given by the recurring images of them as surrounded by water. For instance, before embarking on the journey Dr Real wonders whether to travel by water or land and considers the possible effects of the different journeys on the patients. He finally decides to cross the desert, but 'el río, cuya compañía habíamos rechazado, saliendo de su lecho, vendría a buscarnos por propia iniciativa para imponernos sus leyes rigurosas' [the river, whose company we had rejected, leaving behind its bed, would come to get us by its own initiative to impose on us its rigorous laws].[46] The power of the swollen river is so large that 'sumergía [las tierras] de golpe, arrasando todo a su paso' [it suddenly submerged the lands, devastating everything in its passing] and 'fue ganando las tierras en dirección oeste hasta cubrir la llanura, por mucho que un observador se desplazara en ella a pie o a caballo, en todo el horizonte visible' [it had conquered the land towards the west even to the point of covering the plain, no matter how much an observer might travel through it on foot or on horse, to the whole visible horizon].[47] This characterization of the water (invading the landscape, full of animosity, harrying the travellers), finds justification in what Sarlo has read as one of the trip's main qualities in *Las nubes*:

45 Homi Bhabha, 'The Commitment to Theory', in *The Location of Culture* (New York: Routledge, 2008), 28–56 (53).

46 Saer, *Las nubes*, 88.

47 Saer, *Las nubes*, 113.

> La caravana por el desierto tiene algo de barco atravesando el océano (desde los viajeros son clásicas las comparaciones de la pampa con el mar) [...] además, el territorio que atraviesan está en un estado casi líquido, no sólo por las inundaciones sino por su carácter socialmente informe, la ausencia de instituciones y la precariedad.
>
> [The caravan through the desert has something of a ship crossing the ocean (since the first travellers, the comparisons between the Pampa and the sea are typical) [...] also, the territory they cross is in an almost liquid state, not only because of the floods but also because of its shapeless social character, the absence of institutions, and the very precariousness.][48]

Sarlo's observation on the absence of institutions and the 'shapeless social character' reinforces my points above on the composition of spaces in *Las nubes* as having blurred boundaries and on the madmen as challenging homogenizing national projects. Though Saer accomplishes this by using the very same analogies between 'la llanura' [the plain] and the ocean, which gripped the imagination of travellers in the nascent modern Argentina, he repurposes those images to question both the fixity of boundaries and the processes of homogeneization.

In *Los viajeros ingleses y la emergencia de la literatura argentina*, Adolfo Prieto highlights how certain foundational texts of Argentine literature, such as Sarmiento's *Facundo*, use the analogy of the desert as an ocean, drawing directly on the work of several English travellers (such as Francis Bond Head, Joseph Andrews, and Edmond Temple).[49] *Las nubes* takes the imagery further, describing water as pervading everything also in the form of rain and of giving the impression of coming 'más allá de los sentidos y del pensamiento, de un lugar improbable, exterior al espacio mismo que ocupaba' [from an improbable place, exterior to the very space it occupied, beyond the senses and thought].[50] In this way, like Saer's description of the desert as a repeating landscape, these images of all-pervasive water suggest boundlessness; both water and desert are shown expanding over their own

48 Beatriz Sarlo, 'Aventuras de un médico filósofo', in *Escritos*, 299.

49 Adolfo Prieto, *Los viajeros ingleses y la emergencia de la literatura argentina 1820–1850* (Buenos Aires: Sudamericana, 1996).

50 Saer, *Las nubes*, 120.

horizons and this is also, once again, quite cinematic in the sense that our angle of vision expands with them.

The representation of the desert as full of water offers another related angle from which to read *Las nubes* as connected to the common Renaissance theme of the Ship of Fools. In his work 'La nave de los locos de Juan José Saer' Hugo Vezzetti makes a direct link between *Las nubes* and the Ship of Fools, citing Foucault's in-depth study of the subject.[51] Vezzetti compares how Saer and Foucault explore the symbolic dimensions of both the travellers and the trip. In his view, the travellers's symbolic madness represent how love, knowledge, language and power can be misplaced. Vezzetti considers that the trip in *Las nubes* places insanity on the boundaries of reason, or better, madness as the point from which to investigate the characteristics of mental sanity.[52] In a similar vein, Foucault also sees the Renaissance Ships of Fools as 'highly symbolic', representing 'madmen in search of their reason'.[53] The association of madness and water, which according to Foucault 'have long been linked in the dreams of European man' is an idea that *Las nubes* clearly explores through the caravan of madmen crossing a landscape that is ambiguously a desert and yet full of water.[54] Foucault explains that the motif of the Ship of Fools appears time and again during the Renaissance in literary and pictorial compositions, such as Hieronymus Bosch's paintings *The Cure of Madnes*' (c. 1490) and *The Ship of Fools* '(c. 1500), Sebastian Brant's *Stultifera Navis* (1494), and Josse Bade's *Stultiferae Naviculae Scaphae Fatuarum Mulierum* (1498) and according to Foucault these ships were not merely a product of the Renaissance imagination, they did in fact exist. Found, for the most part in Germany, these real Ships of Fools were used for several purposes: to expel insane people or for the madmen's segregation and cure and as pilgrimage

51 Michel Foucault, *Madness and Civilization: A History of Insanity in the Age of Reason* (New York: Vintage Books, 1988).

52 Hugo Vezzetti, 'La nave de los locos de Juan José Saer', *Punto de vista*, 59 (1997), 39–41.

53 Michel Foucault, *Madness and Civilization: A History of Insanity in the Age of Reason* (New York: Vintage Books, 1988), 9.

54 Foucault, *Madness and Civilization*, 12.

boats. What is more important, however, is that driving the madmen away did 'not assume their entire significance on the plane of social utility or security. Other meanings much closer to rite are certainly present'.[55] For example, in Foucault's view, these trips transformed the madman into a:

> Prisoner of his own departure [...] navigation delivers man to the uncertainty of fate; on water, each of us is in the hands of his own destiny [...] The madman's voyage is at once a rigorous division and an absolute Passage. In one sense, it simply develops, across a half-real, half imaginary geography, the madman's *liminal* position on the horizon [...] his exclusion must enclose him; if he cannot and must not have another *prison* than the *threshold* itself, he is kept at the point of passage. He is put in the interior of the exterior, and inversely. A highly symbolic position, which will doubtless remain his until our own day, if we are willing to admit that what was formerly a visible fortress of order has now become the castle of our conscience.[56]

These words resonate with *Las nubes*' emphasis on the blurred liminal spaces that the characters occupy, for Foucault sees the 'fools' as permanently in the process of passage. Simultaneously, this idea of being caught at a threshold, mainly by the effects of the trip and water, is another theme that *Las nubes* explores in some depth. For instance, when the main character goes to collect one of the patients in order to take him to the asylum, he stays at the patient's parents' house and they get trapped in the city, surrounded by flooded rivers: 'Acorralados por la inundación como estábamos, también la prisión del mundo, reforzada por ese cerco de agua, férrea, se duplicaba' [Surrounded by the flood as we were, also the prison of the world, reinforced by that rim of water, steely, was duplicating].[57] In the novel it is made clear that water is all-pervading, it spreads to cover all the available space, to a point where there appears to be no way out. Likewise, Foucault's description of the functions fulfilled by the Ship of Fools highlights another important feature of Saer's novel, that the presence of water or navigation makes more palpable the uncertainty of destinies. Similarly, a crucial fact noted by Foucault is that the theme of insanity increasingly

55 Foucault, *Madness and Civilization*, 10.
56 Foucault, *Madness and Civilization*, 11 (emphasis in the original).
57 Saer, *Las nubes*, 116.

appeared at the end of the Middle Ages because it articulated the unease and changes that were taking place in European culture at that time. From then on, madness has 'haunted the imagination of Western man'.[58] So we could conjecture about the significance of *Las nubes*' caravan of fools by tracing a connection with Foucault's work, drawing on the fact that both crews were travelling during convulsive times or important moments of radical change.

The world of *Las nubes* seems to be in a constant state of flux. This is noticeable not only in the shifting immensity of the desert/sea, but also in the constant changes produced by the journey itself, including changes in values and norms amongst the travellers. This last point is evidenced most strongly in the reflections of the protagonist at the end of his account. He recalls how the trip transformed the attitudes of the whole caravan, particularly how it released them all from the strictness of conventional morals, such that some of the men started to walk around in loincloths, some of the women even totally naked, and how some of them engaged in undifferentiated sexual relations. Thus, the voyage in *Las nubes*, as Dennis Porter has characterized most types of travel, 'embodies powerful transgressive impulses', that is, it not only challenges conventional values but also overturns them.[59] And while the journey undermines previous principles in *Las nubes*, it also opens up new paradigms or perspectives for the crew: 'El viaje [...] nos había incitado a crear nuestras propias normas de vida [...] a crear un universo exclusivo, más y más diferente a medida que pasaba el tiempo' [The trip had incited us to create our own norms of life [...] to create an exclusive universe, more and more different as time went by].[60] And while as Eric Leed points out, 'the transformation of social being in travel, becoming someone else through territorial passage, is a cliché of the literature,'[61] what is interesting about the ways in which *Las nubes* conceives the transformations of travel is that both individual and social identity are

58 Foucault, *Madness and Civilization*,15.

59 Dennis Porter, *Haunted Journeys* (Princeton, NJ: Princeton University Press, 1991), 9.

60 Saer, *Las nubes*, 191.

61 Eric Leed, *The Mind of the Traveler: From Gilgamesh to Global Turism* (New York: Basic Books, 1991), 263.

posited in fluid terms. Moreover, the strong influence exerted by the journey and the territories traversed on the conduct and morals of the characters defies the reductionism of the exclusionary binary nature-culture, for the behaviour of the 'cultured' (the doctors, soldiers, etc.) is, at least in part, swayed by the effects of nature. Thus, the changes produced by the journey in *Las nubes* also suggest a move beyond long-standing defining polarities.

In many respects, Saer's novel offers the view of a journey and spaces as perpetually changing, engaging the reader with issues of transformation and movement. But if the novel foregrounds the reading of its spatial configurations and trajectories, it does so by unsettling both conceptual and physical limits. *Writing travel* and space in *Las nubes* shows a transgressive impulse, questioning traditional views and emphasizing the difficulties of considering literature in terms of fixed boundaries. Travel is used in *Las nubes* as a metonymy for the process of change and of the overcoming of constrictive antinomies such as civilization-barbarism, and madness and sanity. Similarly to the interaction of the cinematic and the literary in the novel, the 'spatial geographies' of *Las nubes* seem to be generating cartographies that cannot map boundaries, those of the desert, clouds, sea or the nation, but rather sketch their crossings, expansions and modifications. Saer's text reveals how the desert, which was previously defined as the land of barbarism and incorporated as a national space, can be flexibly remodelled and those defining premises questioned. On the one hand, what might seem a fundamental irony at stake in this novel is that Saer's transgressions are dependent on the very 'national' letters on which Argentine literature has been based. On the other hand, any transgression implies this logical disjunction, to challenge specific ideas one has to address them and in one way or another depend on them. The novel not only re-evaluates the place of the nation as a clear and distinct model in literary criticism; the journey and spaces of *Las nubes* force the reader to re-think the continuously evolving confines and encounters that inform cultures.

CHAPTER 2

Stray Journeys: Travel in Roberto Bolaño's *Los detectives salvajes*

Bolaño's novel *Los detectives salvajes* (1998) has been read as deeply affiliated with Mexico, interacting with its culture and literature and retrieving a sense of its terrain and appraising the country.[1] As such, it was defined as 'La más persuasiva de las novelas mexicanas' [The most persuasive of the Mexican novels], and 'acaso la novela más importante que un extranjero ha escrito sobre este país [México] desde *Bajo el volcán*' [perhaps the most important novel that a foreigner has written about this country [Mexico] since *Under the Volcano*].[2] In like manner, Juan Villoro states that *Los detectives salvajes* 'recupera un país único y espectral. En este sentido estamos ante una de las más brillantes novelas mexicanas' [recovers a unique and spectral country. In this sense, we are in the presence of one of the most brilliant Mexican novels].[3] Philip Derbyshire notes that from the opening date of the novel, 2 November (which marks the famous Mexican celebration of the Day of the Dead), 'the most obvious deduction is that this is Bolaño's 'Mexican' novel, that is, it is the novel where he elaborates a judgement on Mexico, the place of his own exile from Chile'.[4] Even Bolaño himself in an inter-

1 Roberto Bolaño, *Los detectives salvajes* (Barcelona: Anagrama, 1998). All further references to this novel are taken from this edition. Subsequently all references to the translation are taken from *The Savage Detectives*, trans. Natasha Wimmer (London: Picador, 2007).

2 Christopher Domínguez Michael, 'Bolaño, Roberto (Santiago de Chile 1953, Barcelona, España – 2003)', *Anales de literatura chilena* (2009), 251–256 (252).

3 Juan Villoro, 'El copiloto del Impala', in *Roberto Bolaño: la escritura como tauromaquia*, ed. Celina Manzoni (Buenos Aires: Corregidor, 2006), 77–80 (78).

4 Philip Derbyshire, '*Los Detectives Salvajes*: Line, Loss and the Political', *Journal of Latin American Cultural Studies: Travesia*, Vol. 18, 2–3 (2009), 167–176 (167).

view published in 1999 described *Los detectives salvajes* as 'una novela de impulso mexicano' [a novel of a Mexican impetus].[5] These considerations resonate, to a greater or lesser degree, with an old debate in which the following appraisal by Fredric Jameson occupies a prominent position: 'Third-world texts, even those which are seemingly private [...] necessarily project a political dimension in the form of national allegory'.[6] Jameson considers that 'the primacy of national allegory in third-world culture' is the only possible response that countries in Africa, Asia and Latin America can offer, after their histories on the receiving end of Imperialism. Conversely, difficulties emerge when carrying out such a reading on 'first-world' literature, for 'in the west, conventionally, political commitment is recontained and psychologized or subjectivized by way of [a] public-private split' whereas for Jameson 'third-world national allegories are conscious and overt'.[7]

Despite the obvious fault with Jameson's statement that for the entire literary production of countries with a history of imperial domination 'the only possible response is a nationalist one',[8] his arguments, the aforementioned readings of Bolaño's novel, and the discussions described in the following paragraphs are a visible crystallization of the controversies that surround the category of the national in relation to literature. To be sure, those discussions also correlate with the most recent geopolitical, conflicting reconfigurations of the local and the global. In this regard, García Canclini's question of how Latin American societies react to the sudden reordering of the last fifty years, whereby 'countries de-industrialize, national democratic institutions weaken, and economic and cultural dependency on the centers of global power intensifies' is pertinent.[9] Though this is not the place to delve into the effects of late capitalism in Latin American countries in a

5 Roberto Bolaño, interview with Ana Anabitarte, *El Universal*, 5 July 1999, 4.

6 Fredric Jameson, 'Third-World Literature in the Era of Multinational Capitalism', *Social Text*, 15 (1986), 65–88 (69).

7 Fredric Jameson, 'Third-World Literature', 71 and 80 respectively.

8 Aijaz Ahmad, 'Jameson's Rhetoric of Otherness and the "National Allegory"', *Social Text*, 17 (1987), 3–25 (8).

9 Néstor García Canclini, *Imagined Globalization*, trans. George Yúdice (Durham, NC and London: Duke University Press, 2014), 4.

systematic manner, the arguments below will, however, help to lend weight to the position I believe *Los detectives salvajes* occupies in relation to the view that 'Third World' literature is necessarily reducible to a national allegory. Accordingly, one of the aims of this chapter is to explore the resonances of the debate on national literatures in *Los detectives salvajes.* A particular focus will be the ways in which Bolaño's novel negotiates its way through a radically contingent reality, engaging with, but also questioning and, at times, ridiculing the validity and persistence of national and continental readings and scope for literature. In this way, a main point of this chapter is that, notwithstanding the frequency of national readings of this novel, and the references to Mexico that appear in *Los detectives salvajes*, an examination of the central tropes of this text permits one to see how, in fact, it does not 'project a political dimension in the form of a national allegory'. As in the previous chapter on Saer's *Las nubes*, I would like to argue here that *Los detectives salvajes* is organized around an outlook created by *writing travel* that challenges the possibilities of assuming a clear and direct association between literature and nation. I also propose that through journeying and the novel's revisionings of space and destination, Bolaño's text reaffirms Jean Franco's appraisal, that from the late 1980s 'concepts of national and Latin American identity started to be inapplicable.[10] In fact, in what follows, I show how neither the spatializing operations of the novel, nor the contested and distanced dealings between the characters and the travelled territories do in fact underpin views of the nation or national attachments.

In response to the scholars' views outlined above, however, we should acknowledge that *Los detectives salvajes* registers the twists and turns of a particular Latin American community in the seventies. What we should ask ourselves, in the manner of Aijaz Ahmad in reply to Jameson's article, is if community and nation are 'the same thing' and whether it is possible for Latin American literature to address a move in which 'one may indeed connect personal experience to a "collectivity" – in terms of class, gender, caste, religious community [...] combining the private and the public, and

10 Jean Franco, *Decline and Fall of the Lettered City: Latin America in the Cold War* (Cambridge, MA: Harvard University Press, 2002), 16.

in some sense "allegorizing" the individual experience without involving the category of the "nation"'.[11]

It is my contention that, in its dealings with the experiences of a community, *Los detectives salvajes* reassesses the poetics of the nation, reformulating what Beatriz Sarlo calls 'La idea clásica de sociedad' [the classic idea of society], which presupposes 'una escena única y nacional' [one and only national scene],[12] while engaging with a 'less restricting idea of collectivity' than that of the nation.[13] Sarlo explains that modern societies were consolidated through processes that unified diverse territories and cultures through the elimination of their differences or 'resistencias' [resistances].[14] In keeping with Ahmad's appraisal, Sarlo adds that some effects of capitalism, globalization and deterritorialization, have transformed the bonds within modern societies and created new relationships more connected to 'escenarios reducidos, menos universalistas y más específicos culturalmente' [limited scenes, less universalistic and culturally more specific].[15] On a related note, Josefina Ludmer remarks that from the 1990s onwards 'la literatura ya no es manifestación de identidad nacional. Se trata de una forma de territorialización que es el sitio y escenario de otras subjetividades' [literature is no longer a national manifestation, it is about a territorialization that works as the scene and place of different subjectivities].[16] As my analysis in this chapter demonstrates, Roberto Bolaño's *Los detectives salvajes* registers a departure from the cultural national models referred to by Sarlo and Ludmer, while exploring the experiences and responses of more discrete communities. In this sense, I do not see *Los detectives salvajes* as trying to contribute to a national cultural capital which was, as Mary Louise Pratt argues, a common concern of early twentieth-century artists

11 Aijaz Ahmad, 'Jameson's Rhetoric of Otherness and the "National Allegory"', *Social Text*, 17 (1987), 3–25 (15).
12 Beatriz Sarlo, *Tiempo presente: notas sobre el cambio de una cultura* (Buenos Aires: Siglo veintiuno editores, 2001), 56.
13 Ahmad, 'Jameson's Rhetoric of Otherness', 15.
14 Sarlo, *Tiempo presente*, 56.
15 Sarlo, *Tiempo presente*, 57–58.
16 Josefina Ludmer, *Aquí América Latina: una especulación* (Buenos Aires: Eterna Cadencia, 2010), 135.

in Latin America.[17] Rather, the impulse of the narrative of *Los detectives salvajes* reacts to the social and cultural changes produced in the second half of the twentieth century by distancing itself from nation-building responsibilities.

As hinted above, this rift from the contribution to a national cultural capital is, in fact, contested and thus the possibility of backing a national literature is indeed considered in the novel, though this project is presented by Bolaño in a humorous manner and as notably ridiculous. For instance, in the second section, one of the character-narrators has a conversation with two of the main characters, Lima and Belano, in which they tell him the reasons why they embarked on an arduous and dangerous journey to find a lost poet and her work: 'lo hacemos por México, por Latinoamérica, por el Tercer Mundo, por nuestras novias, porque tenemos ganas de hacerlo' [we are doing it for Mexico, for Latin America, for the Third World, for our girlfriends, because we feel like doing it].[18] Their reasons, for the hell of it, girlfriends, Mexico and Latin America, all together and with equal importance, as reasons to endure a difficult and dangerous journey, are a waggish response in the novel. Moreover, there is a clear disparity between the import given to their 'national' and 'continental' quest and to its aim since, despite these characters' humorous explanation, it is never clear why finding this poet and her work is important at all.

Bolaño's considerations over furnishing a Latin American cultural project are given an equally jocose treatment. We find a clear example of this in the first section of the novel where the main character explains that the avant-garde movement he has joined expects to 'crear un movimiento a escala latinoamericana' [create a movement on a Latin American scale], to which his interlocutor responds: '¿A escala latinoamericana? No me hagas reír' [On a Latin American scale? Please].[19] These examples are not proof that Bolaño is positing an end to nationalism or that his work is a celebration of postnational realities, but that the position of the writer in

17 Mary Louise Pratt, *Imperial Eyes: Travel Writing and Transculturation* (New York: Routledge, 2008), 230.

18 Bolaño, *Los detectives*, 553, 533.

19 Bolaño, *Los detectives*, 36, 25.

Latin America as the contributor to a national cultural capital has been thrown into crisis in his work, as has the idea of the nation as an explanatory cultural form.

My analysis of *writing travel* in *Los detectives salvajes* not only identifies a distance from the national in Bolaño's novel, it also investigates how writing the journey encapsulates a particular way of seeing texts, how it becomes an optic through which to examine not only space, but also reading, and how to treat literature. Thus, the journey in *Los detectives salvajes* is articulated as a fictional device from which we can take interpretative tools that tend to be consonant with movement; as such, they are visual, mutable and boastful of literature's ephemerality. Therefore, in this chapter I aim to analyse how Bolaño's novel provides a type of literature that seems, in a way, to move against its own pretentions: while it provides stable samples and reflections on 'Mexican' literature, it simultaneously creates a poetics of *writing travel* directed towards a revision and critique of that very view, of an unwavering literary and cultural project.

The writing of travel

The significance of my points above manifests itself in the varied and important ways in which *Los detectives salvajes* is about travel. From the point of view of theme, the ending of the first and third parts narrate the journey of four young Latin Americans from Mexico City to the north of the country into the Sonoran Desert. In the second part of the novel it comes to the reader's attention that two members of those who started travelling in the first section of the novel continued with their journeys through different countries in Europe, and in Liberia, Nicaragua and Israel. Metaphorically, the first part of the novel is also about travel as it is centred on the initiation into poetry of Juan García Madero, a seventeen-year-old orphan who is the main character and only narrator of the first section of *Los detectives salvajes*. His entrance into the city's literary life is marked by his admittance into an avant-garde group, a moment which is described in the opening sentence of the first section of the novel: 'He sido cordialmente invitado a formar parte del realismo visceral. Por supuesto, he aceptado. No hubo ceremonia de iniciación' [I've been cordially invited to join the visceral realists. I accepted, of course. There was no initiation ceremony].[20] Although without ceremony, this initiation begins García Madero's literary *bildungsroman*, a coming-of-age story which is mixed up, in turn, with his initiation into sexual and amorous life. Formally, the presence of images such as windows, zigzags, curved and straight lines, and other drawings not only indicate a play on interpretation, but also that in this novel the configurations of space, in this case the space of the page, are significant. The novel also makes use of different genres, such as the adventure novel, autobiography, road narrative, and crime fiction. None is clearly differentiated from the other, nor are they strictly adhered to as most are transgressed in one way or another. This 'swerving' of genres speaks to the novel's own journeys with form.

In addition, the second part of the novel, divided itself into different fragments narrated by varied characters, has been characterized by Bolaño himself as replicating the flow of a river: thus, he sees in this novel 'una lectura, una más de las tantas que se han hecho en la estela del *Huckleberry Finn* de Mark Twain; el Mississippi de *Los detectives* es el flujo de voces de la segunda parte de la novela' [a reading, one more of the many that have been done in the wake of Mark Twain's *Huckleberry Finn*; the Mississippi of *The Detectives* is the flow of voices of the second part of the novel].[21] Therefore, the journey of Huckleberry Finn in a raft down the river is understood by Bolaño as replicated formally in the variety of voices and fragments of the second part of the novel. Thus, the implications of the fundamental ways in which this novel portrays travel point to what, in my view, is an important and yet understudied subject of the novel: how the journey in *Los detectives salvajes* redefines ways of viewing both literature and the nation. From this perspective, a main aim of this chapter is to analyse representational practices in *Los detectives salvajes* which evidence the novel's mirroring of reading practices

20 Bolaño, *Los detectives*, 13, 3.

21 Roberto Bolaño, 'Acerca de *Los detectives salvajes*', in *Entre paréntesis* (Barcelona: Anagrama, 2006), 326–327 (327).

and ways of travelling. In order to delve into the intricacies of *writing travel* in *Los detectives salvajes*, I offer first, in the following paragraphs, a brief exposition of the novel's spatial tracings and plot.

Los detectives salvajes commences with a diary entry from 2 November 1975. The voice of the journal belongs to Juan García Madero, a Mexican law student dropout, who is trying to become a poet. The date is not random since it coincides with the arrival of Malcolm Lowry (author of the epigraph to *Los detectives salvajes*), in Mexico thirty-nine years earlier. Lowry's arrival in Mexico and the years he spent there were instructive to the writing of *Under the Volcano* (1947), a highly autobiographical work that Lowry was to conclude almost a decade after he set foot in the country. The epigraph reads: '¿Quiere usted la salvación de México? ¿Quiere que Cristo sea nuestro rey? –No' [Do you want Mexico to be saved? Do you want Christ to be our King? 'No'], thus placing the action in Mexico and, together with the first diary entry, bestowing a referential tone upon the novel and its setting. Another important point is that the date is also that of the Day of the Dead, a famous national celebration in Mexico. In stark contrast to Lowry's famous novel, which is highly descriptive, *Los detectives salvajes* lacks detailed descriptions of settings. The absence of iconography or detailed visual images of the city in Bolaño's novel steers it clear of connections between its characters' metropolitan experience and that of the epitomic roving figure of the modern city: the flâneur. In this regard, my view differs from that of Fernando Saucedo Lastra, who does indeed characterize the main protagonists of the first part of the novel as that prototypical stroller.[22] In my view, the characters of *Los detectives* are representative of urban dwellers whose lives tally more with Franco Moretti's observation that 'city life does not allot time to contemplation'.[23]

What we read in this first part of the novel is, moreover, not a totalizing view of the city but rather a particular experience of life in the streets

22 Fernando Saucedo Lastra, 'El país enemigo: México en la obra de Roberto Bolaño 1980–2004' (unpublished doctoral thesis, Universitat Pompeu Fabra, 2012), 108 <http:hdl.handle.net/10230/19805> [accessed 4 June 2013].

23 Franco Moretti, *Signs Taken for Wonders: On the Sociology of the Literary Forms* (New York and London: Verso, 2005), 125.

and houses, generally guided by a sense of randomness of movement and a loss of connection with the particularities of the place. The references to Mexico City are superficial and made in passing, the names of a few streets and emblematic bars are mentioned, but there is no further engagement with them. For instance, García Madero's diary entry of 7 November opens with the statement that 'La Ciudad de México tiene catorce millones de habitantes' [There are fourteen million people living in Mexico City],[24] but this does not generate a description of the inhabitants of the city or even the individual experience in that space amongst that number of people. Instead, the following sentence just states 'No volveré a ver a los real visceralistas' [I'll never see the visceral realists again] and continues by offering some details of the books and poems García Madero has read.[25] Though that entry situates the action in a place and time, it shows that the view of Mexico City in *Los detectives salvajes* is not illustrated by topography or local detail. The Mexican spaces are experienced through aesthetic detachment, and personal and internal experience in movement are prioritized over an engagement with the physical world: the title of the first section of the novel, entitled 'Mexicanos *perdidos* en México (1975)' [Mexicans *Lost* in Mexico] is symptomatic in this regard.[26]

In the first section we also read about the life of Juan García Madero and the events he goes through from the moment he joins a group of avant-garde poets called the 'realvisceralistas' [visceral realists]. His diary takes the reader through García Madero's sexual experiences, the quotidian problems of living together, first with his uncle and aunt and then Rosario (a young bartender who takes him in to live in her rented room), literary discussions, love affairs, stories of friendship – in sum, all the minor details of the life of a group of vibrant and young Mexican writers. At some point in his diary we see García Madero becoming especially involved with María Font, a young poet and artist whose father, Joaquín Font, designs the magazine for the visceral realists. Joaquín Font in turn has an affair with Lupe, a prostitute friend of his daughter who wants to quit her job as a sex worker and start

24 Bolaño, *Los detectives*, 21, 11.
25 Bolaño, *Los detectives*, 21, 11.
26 Emphasis mine.

a new life and whom he eventually takes in to live with his family. The ups and downs of the lives of the two main leaders of the visceral realists, Arturo Belano and Ulises Lima, are commented upon in García Madero's diary, as young poets who live off selling marihuana and doing menial jobs, interrupt poetry gatherings, attend parties and travel (these two characters come to the foreground in the second part of the novel, as the main subjects spoken of by various narrators). García Madero's diary starts to draw to a close when he, Ulises Lima, Arturo Belano, Lupe and other members of the visceral realists are confined to the Fonts' house besieged by Lupe's irate pimp and his sidekick – who demand her return. The diary breaks off when Lima, Belano, Madero and Lupe make a narrow escape in Joaquín Font's new Ford Impala towards the north of Mexico. Their escape is thus suggestive of contemporary transformations of urban mobility that, as Edna Duffy explains, moved from: 'Flânerie, the phenomenon of the pedestrian wandering the city pavements' which 'had been a rich motif of nineteenth-century urban culture', to an abandoning of that figure 'and unceremoniously plump[ing] it into public transport or his new car'.[27]

While the journey in *Los detectives salvajes* is obviously triggered by Lupe's need to flee from her pimp, the travellers also have other additional objectives in mind. Parallel to their escape, the passengers of the Impala aim to start on a quest to find a poet called Cesárea Tinajero and her complete literary work. Lima and Belano are peculiarly enthralled with the idea of finding this poet, who was one of the forerunners of an earlier group of visceral realists in the 1920s (from whom Belano and Lima borrowed the name). Tinajero's work, although seemingly known to other writers of that time, is nowhere to be found. She also seemed to have disappeared years before into the north of the country, in the Sonoran Desert. Thus, although the journey is primarily a getaway, it cannot be separated from the literary quest and literary motifs within which the narrative is inescapably caught.

García Madero's diary is brought to a standstill as the reader enters the second section of the novel, entitled 'Los detectives salvajes (1976–1996)'. The following 400 pages feature the testimonies of an array of new narrators:

27 Edna Duffy, *The Speed Handbook: Velocity, Pleasure and Modernism* (Durham, NC: Duke University Press, 2009), 66.

all acquaintances, friends, bosses, lovers or rivals of Belano and Lima. These fragments are all narrated in the first person and give personal accounts of events and experiences in the relaters' lives as they intersect with the two visceral realists. Some testimonies give contradictory information about Lima and Belano; leaving to the reader to decide how trustworthy each narrator is and how the testimonies should be interpreted. In most cases the voices talk to an unidentified person who seems to be interviewing them and, on a few occasions, they seem to be addressed to Belano as if he were conversing with them himself. Compared to the early pages of the novel, the text here takes on a stronger oral quality.[28] The stories told by these narrators are always introduced with a journalistic dateline specifying the exact time and place of each conversation.

The varied monologues take place at diverse and random locations such as a mental hospital on a particular highway in Mexico, parks in Mexico City, bars in the US, a university campus, a prison and a house in Tel Aviv. Through the accounts given by the different narrators we can piece together several possibilities for Belano and Lima's itineraries which span several countries including Mexico, Nicaragua, and Angola and include stops in cities such as Paris, Barcelona, and Tel Aviv. Although some of these testimonies contradict each other, we do gather some information (if dubious) about the numerous affairs these two characters may have been involved in during the years covered in this part of the novel. It is said that Belano and Lima travel separately and live in different cities. We read about their many and varied jobs which include a security guard at a camp site, a bar tender, dishwasher and fisherman. We also learn how they try to continue former romantic relationships and even serve a sentence in an Israeli prison and fight a duel with a literary critic over an apocryphal poor review of one of Belano's novels. The multiple narrations point to a representation of these characters' lives and journeys as varied but also incomplete and distorted. The different locations, Latin American accents, and portrayals of Belano

28 For more on this, see Marco Kunz's 'Cuando Bolaño "habla" mexicano, en torno a la oralidad fingida en *Los detectives salvajes*', in *Roberto Bolaño. Estrella cercana. Ensayos sobre su obra*, eds Augusta López Bernasocchi and José Manuel López de Abiada (Madrid: Verbum, 2012), 148–175.

and Lima within these accounts give a fluency of scene, characterization and relationships. Likewise, national and regional borders, both geographical and linguistic, are crossed with such frequency that the notions of home and away seem to continuously branch off and lose relevance.

The third and last section, entitled 'The deserts of Sonora (1976)', reverts to the Sonoran Desert and to the fate of the four fugitives in the Impala. This part of the novel is also narrated through the diary and voice of García Madero. In this section the reasons for Lima and Belano's subsequent peregrinations (narrated in the second section of the novel) are brought into relief. The reader also, finally, discovers the whereabouts of Cesárea Tinajero and her work, although a few issues are not finalized or remain unsaid, such as what happens with Belano in Africa and with García Madero and Lupe in the north of Mexico. These omissions give the ending a quality of elusiveness and openness. The end of the novel pictures an empty window with the question: 'What's outside the window?', and therefore lures the reader into imagining a continuation of the journey. Thus, travel, in this novel, does not trace a path away from home and back again, and there are no guaranties of reaching a destination or of a return. In this way, opposing paradigms seem to define the journey: one that could be called transcendental, in which a literary gain is pursued – to get access to Tinajero's work – and which moves them beyond ordinary experience, and another, an open-ended escape more typical of a thriller and devoid of trascendentality.

The final window in *Los detectives salvajes* could also be read as a riposte to the possibilities of fulfilment of the journey, a sheer emptiness, and hence a questioning of the validity and extent of the journey as transcendental. However, as Philip Derbyshire rightly points out, the final window can be seen in connection with the epigraph of the novel, based on the fact that *Los detectives salvajes* 'moves from the question of action implied in the quotation from Lowry to a metaphysical question: the novel ends with a question and a non-answer'.[29] Thus, the non-answer constituted by the open window is seen by Derbyshire not as undermining, but as reinstating, the journey's transcendental value. Ultimately, this open window (portrayed

29 Derbyshire, '*Los Detectives Salvajes*: Line, Loss and the Political', 167.

as an 'end' to both the journey and the novel) highlights the very refusal of a conclusive ending. The possibilities of achieving a goal, a specific meaning and of denouement are recast in favour of an experience of place and destination as both uncertain and continuous.

Perceptions of automobile journeys

In the first section of the novel, before the escape, there is one particular instance in which Mexico City is briefly described by García Madero as he sees it from the back seat of a taxi. Fleeting images of the metropolis are observed while framed by the taxi's window. These compose an almost televisual display that casts the character as a passive spectator and drive him to consider their lives in the city as shut up in solitude:

> Y yo aproveché para contemplar el paisaje que se sucedía monótono en la ventanilla: las fachadas de la Juárez y de la Roma Norte, y también me puse a pensar en María y en lo que me separaba de ella [...] y me puse a pensar en Rosario y en nuestro cuarto de vecindad [...] Y también me puse a pensar en mis tíos e incluso me pareció verlos, alejándose por una de aquellas calles por las que pasábamos, tomados del brazo, sin volverse para mirar el taxi que se perdía zigzagueando peligrosamente por otras calles, inmersos en su soledad así como Pancho, el taxista y yo íbamos inmersos en la nuestra.
>
> [I stared out of the window, watching the scenery (the storefronts of Juárez and Roma Norte) rolling monotonously past, and I also started to think about María and what separated me from her [...] and about Rosario and our tenement room [...] And I also started to think about my aunt and my uncle and I even thought I saw them, walking arm in arm down one of the streets that we were passing, never turning to look at the taxi as it zigzagged perilously away down other streets, the two of them immersed in their solitude just as Pancho, the taxi driver, and I were immersed in ours.][30]

It is clearer in the Spanish version of the passage quoted above than in the English translation that the landscape streams by monotonously in the

30 Bolaño, *Los detectives*, 111–123.

window. Thus, the 'paisaje' [landscape] is characterized as 'monótono' as a result of the speed of the taxi which produces an effacement of details. This description of García Madero's taxi itinerary foreshadows the pre-eminence of the road-trip style of the third part of the novel, since the display of images as framed by the window of the car, and the vehicle as a source of monotony, are customary themes of road narratives.[31] García Madero's encapsulation in the car and vision of the window as a screen, incite his perceptual passivity. Moreover, the isolation created by the car drives this character to introspection, a condition that is reinforced by the length of the sentence after 'me puse a pensar', which sums up several clauses separated just by commas and the conjunction 'y', and gives the impression of reproducing this character's stream of consciousness. García Madero thinks specifically about his relationships with María and Rosario, and of his uncle and aunt, and of him and his friend as 'inmersos en su soledad' [immersed in their solitude]. This type of segregation, that of the city dwellers, is a common characteristic attributed to urban life, which can be, as in this quote, accentuated by technology. Thus, the car not only heightens the separation of the inside from the outside, it also intensifies the individual withdrawal and solitude of each of the characters. In the quoted passage we read that García Madero captures ephemeral glimpses of the 'fachadas' [storefronts] from the back of an accelerating taxi; nevertheless, the characters of *Los detectives salvajes* are not really urban observers. Although they can see the fleeting images of the metropolis, their travelling speed erodes the possibility for inspection or lingering on the streets so as to become strolling 'readers' of those very facades. Therefore, the importance of introducing this image is that, while it recapitulates previous events of the novel, it suggests in advance the road journey the protagonist is going to embark on, and encapsulates what could be called the 'speed of view' of the text, casting García Madero as a fleeting and detached observer of the Mexican geographies he travels through. Likewise, the car acts as a

31 For more on this see, Ronald Primeau, *Romance of the Road: The Literature of the American Highway* (Bowling Green, OH: Bowling Green State University Popular Press, 1996), 86.

technological medium for viewing reality, therefore influencing a view of repetition, monotony, and speed.

The beginning of the journey of the four characters is described at the end of the first section of the novel. The narrator recounts that García Madero acknowledges to himself that he is ready and willing to separate from his current environment, and this feeling arrives as an instantaneous decision of a hitherto unrecognized desire, at the point of hitting the pimp outside the Font's family home. Positioned beside the open car with his friends and Lupe inside, he jumps in and states that in that instant 'supe que siempre había querido marcharme. Entré y antes de que pudiera cerrar Ulises aceleró de golpe' [I realized that I'd always wanted to leave. I got in and before I could close the door Ulises stepped on the gas].[32] Although this will to leave a previously known and partly domestic environment seems sudden, it can be anticipated from García Madero's previous words of rejection for the conventionality of his uncle and aunt's life and for their desires (they want him to study law instead of literature, to come back home at night, and so on) and the detachment he shows for the couple's life he shares with Rosario, the partner he moves in with in a room in the city in the first part of the novel.

The motivations for the journey are not, however, merely the desire to flee from the domestic space or Lupe's pimp, but also, as I mentioned above, related to the literary. The characters can therefore be cast both as runaways and literary 'investigators'. Their journey mixes altruistic intentions (to liberate Lupe from her former life and persecutor), with a literary search and the thrill of running away. In addition, this desire to escape, and the very getaway in the car, relates them to other types of road travellers such as the Beat generation. This was acknowledged by Bolaño himself, who in an interview held in 1998, explained that this group was of great influence in shaping the notion of travel within his own literature.[33] Moreover, one of the narrators of the second part of the novel also notes that both Belano and Lima reminded her of the Beats. Thus, in a conscious move,

32 Bolaño, *Los detectives*, 136–124.

33 Roberto Bolaño, 'De aquí a la eternidad: interview with Sergio Paz', *El mercurio*, 13 December 1998, 4–6.

Bolaño equates the travellers of *Los detectives salvajes* with the members of the Beat generation, a group of writers who interpreted the road as an escape and a vehicle for the rejection of the American way of life and social conformity. For the characters of *Los detectives salvajes*, the lure of travelling at high speed in the Impala towards the north of Mexico, also offers them a hasty and adventurous rift from the ordinary. Patricia Espinosa suggests that the journey in *Los detectives salvajes* is 'casi sin retorno, pero tremendamente heredero de la trascendentalidad impuesta por el camino de Jack Kerouac' [almost without return, but very much heir to the transcendental quality of Jack Kerouac's journey].[34] Although the transcendental nature of travel is commonly connected to antiquity and pilgrimage and, as Jaś Elsner and Joan-Pau Rubiés argue, to 'a traveller whose journey brought inner as well as outer fulfilment, return to spiritual plenitude lost in the travails of life, as well as success in the sense of worldy achievement,'[35] the 'transcendentalidad' of the Beats referred to by Espinosa has been critically regarded as a process of self-discovery and self-definition that occurs with the voyage. In Bolaño's novel this is observable in the character of García Madero, whose initiation into manhood and poetry that takes place from the outset of the novel, is given an even stronger impulse with the start of the journey. Nevertheless, this feature should be seen not only as coming from the Beat generation; it also conforms with a general and contemporary conception of travel that, as Eric Leed highlighted, 'signifies autonomy and is a means for demonstrating that one "really" is independent of one context or set of defining associations'.[36]

Let us now turn to how the description of the departure continues:

> Me volví y a través de la ventana trasera vi una sombra en el medio de la calle. En esa sombra, enmarcada por la ventana estrictamente rectangular del Impala, se concentraba toda la tristeza del mundo [...] nuestro coche daba un salto y dejaba atrás

34 Patricia Espinosa H., 'Roberto Bolaño: un territorio por armar', in *Roberto Bolaño: la escritura como tauromaquia*, ed. Celina Manzoni (Buenos Aires: Corregidor, 2006), 125–132 (126).

35 Jaś Elsner and Joan-Pau Rubiés, *Voyages and Visions: Towards a Cultural History of Travel* (London: Reaktion Books, 1999), 9.

36 Eric J. Leed, *The Mind of the Traveller* (New York: Basic Books, 1991), 13.

> la casa de las hermanas Font, el Camaro de los matones, la calle Colima y en menos de dos segundos ya estábamos en la avenida Oaxaca y nos perdíamos en dirección al norte del DF.
>
> [I turned around and through the back window I saw a shadow in the middle of the street. All the sadness of the world was concentrated in that shadow, framed by the strict rectangle of the Impala's window [...] our car leaped forward and left behind the Fonts' house, the thugs' Camaro, Calle Colima, and in less than two seconds we were on Avenida Oaxaca, heading north out of the city.][37]

The separation from the past ignited by the beginning of the journey is epitomized in García Madero's view as framed by the car's back window. García Madero notes that what is left behind is simultaneously a gloomy shadow and clear awareness of familiar places and people. He marks the departure, and signals the discontinuity from the past, with the gesture of turning around and listing what is left behind. The above passage confirms Leed's idea that 'departure is always a break, both an ending and a beginning: it evokes a past and projects a future'.[38] The future anticipated from the car is described with the verb: 'perderse' (which features prominently and significantly in the novel) and 'en dirección al norte del DF'. I return later in this chapter to how the constant appearance of the verb 'perderse' [to get lost] dialogues ostensibly with form and with views of the national territory. In the passages quoted above, both cars 'get lost', a fact which helps to underpin the idea that the traveller is not participating in a journey of exploration and that narrow observation and empirical examination of the landscape are not features of this journey. Furthermore, travelling, as an aesthetic of 'getting lost', blurs the importance of goals and destinations. The open window presented at the end of the novel also reinforces this last idea and, at the same time, through its openness makes the reader see the fulfilment of arrival, or completion, as constantly deferred.

It remains important to highlight that in the last quoted passage the 'salto' [leap] into the future is given by the car, since it defines their journey as a road narrative and simultaneously makes the travellers participants

37 Bolaño, *Los detectives*, 136–137, 124.
38 Eric J. Leed, *The Mind*, 31.

of particular symbols of rank marked by the very medium they use. As Cynthia Golomb Dettelbach remarks, in her analysis of the automobile in American literature, 'the car takes on values and connotations far beyond its physical function'.[39] The automobile in which Belano, Lima, García Madero and Lupe travel in 1975 is a privileged middle-class vehicle, however, the characters are depicted as unaffiliated to those characteristics and class for they have taken or stolen the car and, at the same time, they are laughably ignorant of cars' properties. In spite of the fact that they remain unaware of the rank potentially bestowed upon them by this car, the Impala becomes an object of deliberation and contemplation for the travellers. For instance, while they are travelling in the Impala, García Madero overhears a conversation between Belano and Lima in which they make naïve comments about the noise and movements of the vehicle that make him realize that 'antes de quedarme profundamente dormido me doy cuenta de que ninguno de los dos sabe nada de automóviles' [before I fall into a deep sleep I realize that neither of them knows anything about cars].[40] Although in road narratives the car adds to travel literature 'its own unique merging of the frontier spirit and the worship of the machine as a complex icon',[41] here the 'icon' is deprived of homage and remains unintelligible to the travellers.

That said, the vehicle is central in terms of plot. For instance, it plays a key part in the visceral realists' rivalry with Alberto, Lupe's pimp, with the persecution of Lupe being concretized in the image of the Camaro chasing the Impala. Furthermore, the confrontation between the two groups (the four travellers in the Impala versus Alberto and his friend) is staged very much as a hostile meeting of cars: 'vimos un coche que venía en sentido contrario. Probablemente aquél y el nuestro eran los dos únicos automóviles en muchos kilómetros a la redonda. Por un segundo pensé que nos íbamos a estrellar' [we saw a car coming from the opposite direction. We were

39 Cynthia Golomb Dettelbach, *In the Driver's Seat: The Automobile in American Literature and Popular Culture* (Westport, CT: Greenwood Press, 1976), 10.

40 Bolaño, *Los detectives*, 568–538.

41 Ronald Primeau, *Romance of the Road: The Literature of the American Highway*, 5.

probably the only two cars for miles around. For a second I thought we were going to collide].[42]

The car also permits one to see clearly what Duffy calls the 'politics of speed' of the text, for it affects the travellers' 'sense of space' as distances are 'foreshortened and space condensed'.[43] In Duffy's view, travelling at high speed corrodes the importance of territoriality while offering a sense of emancipation to the traveller since 'speed was envisioned not only as pleasure but as a measure of extraordinary personal power'.[44] In the following section I analyse precisely how the 'politics of speed' in *Los detectives salvajes* determines not only a way of conceiving and observing geographies, but also ways of reading.

Reading the journey

Before the main characters of the novel embark on their literary quest, the only piece of writing that Belano and Lima have seen by Tinajero is a visual poem called 'Zión' which consists of a straight line, a wavy line and a jagged line, each with a half square on top of them. The value of this poem, and even whether it could be considered literature or not, is a point of discussion in the novel. For instance, Lima, Belano and Amadeo Salvatierra, one of Tinajero's friends and a narrator of the second part, have the following dialogue about the poem (Belano asks the questions and Salvatierra replies):

> ¿Qué significa para ti el poema? Nada, dije, no significa nada. ¿Y por qué dices que es un poema? Pues porque Cesárea lo decía, recordé yo. Por eso y nada más, porque tenía la palabra de Cesárea. Si esa mujer me hubiera dicho que un pedazo de su caca envuelta en una bolsa de la compra era un poema yo me lo hubiera creído. Qué moderno, dijo el chileno.

42 Bolaño, *Los detectives*, 603, 571.
43 Duffy, *The Speed Handbook*, 8.
44 Duffy, *The Speed Handbook*, 8.

> [What does the poem mean to you? Nothing, I said, it doesn't mean a thing. So why do you say it's a poem? Well, because Cesárea said so, I remembered. That's the only reason why, because I had Cesárea's word for it. If that woman had told me that a piece of her shit wrapped in a shopping bag was a poem I would have believed it, I said. How modern, said the Chilean.][45]

Cesárea Tinajero's poem, which could be read 'at speed', dramatizes the act of reading as an activity of 'seeing'. In fact, Amadeo Salvatierra reveals that he has been *looking* at this poem for more than forty years and never understood it. He then asks Belano and Lima to explain the mystery of the poem to him to which they respond that there is really no mystery. There is no pretence of finding an intrinsic sense in the poem, there is no satisfactory solution to it or a legitimate reading or explanation for it. Moreover, what is being foregrounded in the poem are clearly not words, but lines, and the piece does not invite one to an exercise of close reading, but to an act of reading as literally *going over* lines. For Emilio Sauri, it is 'precisely the attempt to undo the distinction between *seeing* and *reading*' that underlines the discussion over the poem's meaning.[46]

I have discussed above how what Duffy calls 'the politics of speed' determines the type of relationship that the traveller establishes with the landscape. Here a similar 'politics of speed' seems to be in operation but, in this case, affecting the literary text, in the sense that its meaning has also been 'foreshortened' into lines and squares. The speed of view of both the journey and the text seems to undermine the relevance of contemplation and exploration, but equally thwarts the possibility of fully apprehending the nature or 'mystery' of both the landscape and poem. It is interesting in this regard that Jorge Volpi characterizes Bolaño's descriptions of Mexico and Chile as 'hologramas de la región: fragmentos truncos y dispersos, movedizos y volátiles' [holograms of the region: truncated and disperse, moving and volatile fragments].[47] Volpi infers from

45 Bolaño, *Los detectives*, 398, 374.

46 Emilio Sauri, '"A la pinche modernidad": Literary Form and the End of History in Roberto Bolaño's *Los detectives salvajes*', *MLN*, Vol. 125, nº 2 (2010), 406–432, 409.

47 Jorge Volpi, *El insomnio de Bolívar: cuatro consideraciones intempestivas sobre América Latina en el siglo XXI* (Buenos Aires: Debate, 2009), 176.

Bolaño's 'hologramatic' portrayal of those countries that his literature exhibits 'un carácter *posnacional*' [a postnational character].[48] While I read Bolaño's encoding of travel and writing as a departure from national and continental readings, I believe that *Los detectives* foregrounds the very uncertainty left by the retreat of those paradigms, rather than declaring their complete demise. The prominence of the verb 'perderse' [get lost] in the novel is a particular case in point in this regard. Likewise, the open window at the end of the novel, condenses that uncertainty both in its openness and lack of words. For this reason, while I concur with Volpi in that Bolaño's descriptions of Mexico and Chile are 'disperse, moving and volatile', I am reluctant to fully support Volpi's assertion of a postnational literature.

Rather than openly and resolutely postnational, I see the narrative of *Los detectives salvajes* as encoding the quandaries of a generation which has seen the hopes of national responses to Latin American realities run their course. Further proof of this is that, even though the main characters only see a small sample of Tinajero's work (and the value of the same is highly questionable), Belano and Lima still embark on a long quest, driving from Mexico City to the north of the country, researching in libraries and archives, visiting different towns and asking the locals questions about Tinajero and her oeuvre. But their quest does not, of course, bring the results they searched for, for they prompt Cesárea Tinajero's death (she is killed by Lupe's pimp almost immediately after they find her); they never get hold of her work; and then they have to continue on the run, not least because they then kill the pimp who, in turns out, had allies within the corrupt police force. In consonance with these points, Juan Antonio Masoliver Ródenas sees the quest in *Los detectives salvajes* as 'la desgarradora búsqueda de una generación [...] en el vacío y que, en un país sin futuro, solo parece encontrar respuesta en el pasado ya perdido' [the heartrending search of a generation [...] in the void and in a country without future that only seems to find an answer in the lost past].[49] Jason Wilson likewise considers

48 Volpi, *El insomnio de Bolívar*, 176.

49 Juan Antonio Masoliver Ródenas 'Espectros Mexicanos', in *Roberto Bolaño: la escritura como tauromaquia*, ed. Celina Manzoni (Buenos Aires: Corregidor, 2006), 65–69 (68).

that *Los detectives salvajes* is 'an elegy for a generation of Latin-American would-be poets' and Belano and Lima seem to give up art 'for something never defined that seems to be willed failure and uprootedness'.[50]

The sense of loss to which both Wilson and Masoliver Ródenas refer may explain the intermittent reference, within Bolaño's novel, to lost pieces of writing such as that of Cesárea Tinajero and to lost writers (such as Ambrose Bierce, Rimbaud, and his own Ulises Lima). In fact, Belano's and Lima's own story is presented very much as a writing that has travelled, got lost, and forgotten. *Los detectives salvajes* thus puts forward a type of literature that cannot reassert itself in national or continental enterprises but that is conscious of its own contingency. For instance, in the second section of the novel one of the narrators, the writer Michel Bulteau, explains that Lima described their lives as:

> Una historia de poetas perdidos y de revistas perdidas y de obras cuya existencia nadie conocía una palabra, en medio de un paisaje que acaso fuera el de California o el de Arizona o el de alguna región mexicana limítrofe con esos estados, una región imaginaria o real, pero desleída por el sol y en un tiempo pasado, olvidado o que al menos aquí, en París, en la década de los setenta, ya no tenía la menor importancia. Una historia en los extramuros de la civilización, le dije.
>
> [A story of lost poets and lost magazines and works no one had ever heard of, in the middle of a landscape that might have been California or Arizona or some Mexican region bordering those states, a real or imaginary place, bleached [*sic*] by the sun and lost in the past, forgotten, or at least no longer of the slightest importance here, in Paris, in the 1970s. A story from the edge of civilization, I said.][51]

What is worth noting in this description is that the indeterminacy is both textual (lost works) and geographical (California or Arizona or Mexico). The region that Lima talks about is, in any case, described as bordering another, in a similar way to how the story itself borders both fact and

50 Jason Wilson 'Epic Elegy for a Lost Generation', *The Independent*, 24 August 2007 <http://www.independent.co.uk/arts-entertainment/books/reviews/the-savage-detectives-by-roberto-bolantildeo-trans-natasha-wimmerbrlast-evenings-on-earth-by-roberto-bolantildeo-trans-chris-andrews-462725.html> [accessed 10 October 2010].

51 Bolaño, *Los detectives*, 240, 221.

fiction: 'una región imaginaria o real'. The Spanish version of the passage describes how the region that Lima speaks of could be either real or imaginary, but certainly ('pero'): 'desleída por el sol'. On the face of it, 'desleír' can be defined as to attenuate the expression of thoughts, ideas or concepts. Here 'desleída' could apply to both the region and the story which Lima is bringing up. Thus, an account that is 'desleída' means one that has been attenuated, in this case by an exterior phenomenon, the sun, whose strength is particularly important in the desert regions where the story may have taken place. In this way, the intensity of the writing or the account diminishes with the journey and, finally, when transported into Paris, 'no tenía la menor importancia', loses all importance. There is, however, another possible interpretation of the verb 'desleír'. Taking other examples of how the common prefix des- is used in Spanish, such as in desarmar [disassemble], destejer [undo], desandar [go back along] it is possible to conclude that the region the passage is talking about had been read (leída) and now un-read by an exterior phenomenon. This signals the act of movement in writing, and a consideration of the act of reading as tracing and then losing that trace.

Not only, however, are writing and reading here elaborated as travel, but the reverse movement also takes place: the positing of travel as a fleeting text. When a friend of Belano meets his sister, she is told about how Belano travelled all over Latin America, went back to Chile in 1973 to support Salvador Allende, and how he suffered in a hostile and dangerous place. The friend then imagines Belano in his voyages as 'perdido en un espacio en blanco, un espacio virginal que poco a poco se iba ensuciando, emborronándose' [lost in a white space, a virgin space that kept getting dirtier and more soiled [*sic*]].[52] The metaphor that appears in this passage of the blank space that gets dirtier and then fuzzy (emborronándose) is clearly suggestive of the space on the page and the transient process of writing.

52 Bolaño, *Los detectives*, 166–167, 151.

A journey of genres

In previous sections I have identified some features of *Los detectives salvajes'* voyage that fit with the genre of travel writing, despite the fact that this genre, as I mentioned in the introduction to this book, is an umbrella term that encompasses a variety of styles and techniques for writing about travel. In fact, scholars such as Jan Borm claim that travel writing is not strictly 'a genre, but a collective term for a variety of texts both predominantly fictional and non-fictional whose main theme is travel'.[53] The affiliation of *Los detectives salvajes* with that 'group of texts' also stems from its links with adventure and travel narratives: The journey of Lima, Belano, Lupe and García Madero to an elsewhere, in this case to the Sonoran Desert by car, as an escape from both Lupe's pimp and domesticity but also as a quest, is a common element of quest novels and road narratives. A differentiating factor, however, is that all of these genres or modes of writing tend to have a circularity that *Los detectives salvajes* lacks. Generally, as Georges Van Den Abbeele explains:

> In order to have an economy of travel, some fixed point of reference must be posited. The economy of travel requires an *oikos* (the Greek for 'home' from which is derived 'economy') in relation to which any wandering can be *comprehended* (enclosed as well as understood). In other words, a home (land) must be posited from which one leaves on the journey and to which one hopes to return.[54]

In *Los detectives salvajes* the journey starts in Mexico City and continues to the north of the country; we don't know where Lupe and García Madero end up, but we know they have no desire to return. Belano and Lima go back to the city but then continue their travels through Latin America, Europe, the Middle East and Africa. While most of the main characters

53 Jan Borm, 'Defining Travel: On the Travel Book, Travel Writing and Terminology', in *Perspectives on Travel Writing*, eds Glenn Hopper and Tim Youngs (Aldershot: Ashgate, 2004), 13–26 (13).

54 Georges Van Den Abbeele, *Travel as a Metaphor: from Montaigne to Rousseau* (Minneapolis: University of Minnesota Press), xxvii–xxviii.

in the novel were born in Mexico, Belano is Chilean and therefore started his journey years before and traversed different places to those travelled in the novel. So, while Bolaño's text seems to follow some of the norms of traditional travel narratives, it does not hew close to the logic of return and the sense of home. That distinction is fundamental in so far as it also distances the novel from readings related to the national. The narrative in *Los detectives salvajes* wanders away and loses touch, not only from the views passing by in front of the car, but from the 'fixed point of reference' referred to by Van Den Abbeele.

Los detectives salvajes's work with genres validates what Derrida calls the 'law of genre', which operates on a 'principle of contamination' and simultaneous exclusion, for though 'every text participates in one or several genres', since 'there is no genreless text: there is always a genre and genres, yet such participation never amounts to belonging'.[55] In fact, Tzvetan Todorov considers that 'this might be the very sign of the authentically modern writer – one who no longer respects the separation of genres'.[56] My interest here, however, is not just to determine in what senses *Los detectives salvajes* bypasses the separation of genres, but also to establish the range of interpretative practices or reflections that emerge from its participation in, and simultaneous transgression of, several genres. In relation to this, in an interview given in 1998, Bolaño explains that 'hay una parte de *Los detectives salvajes* que puede ser considerada como un thriller. Pero luego hay una parte que claramente es una novela río, una novela de iniciación' [there is a part of *Los detectives salvajes* that could be considered as a thriller. But there is another part that is clearly a river novel, a novel of initiation].[57] Despite Bolaño's assertion, however, it is not very clear from the reading of *Los detectives salvajes* which section pertains to the 'thriller' (and Bolaño stops short at explaining which 'part' of the novel could be seen as such),

55 Jacques Derrida, 'The Law of Genre', *Glyph 7* (1980), 202–232 (212).

56 Tzvetan Todorov, 'The Origin of Genres', *New Literary History*, Vol. 8, No. 1 (1976), 159–170.

57 Roberto Bolaño, interview with Gonzalo Núñez, 'Soy un pobre aprendiz', *La hora*, 3 November 1998, 30 (translation mine).

however, the word 'detectives' in the title does set up the expectation of a work that should contain at least a crime and some form of an investigation.

Bolaño seems to differentiate the second part of the novel (*la novela río*) from the thriller, yet, there are aspects of the narrations in this section that seem, potentially, thriller-like. As Susanne Hartwig observes, these narrations of the second part do seem to belong to that genre as 'parecen declaraciones dentro del marco de una encuesta o de un interrogatorio' [they seem to be declarations within the frame of a survey or interrogation].[58] It should be acknowledged, nevertheless, that none of the narrators of the second part of the novel actually gives testimony about a crime. Indeed, there is a striking indecisiveness with regards to what the crime is in *Los detectives salvajes*. It could be the killing of Cesárea Tinajero by the pimp or the killing of the pimp by Belano and Lima, but in either case the investigation carried out by the detectives does not concern any of those crimes, rather it is about finding the whereabouts of a writer, Tinajero, and her work. The assassination of both the pimp and Tinajero take place briefly at the end of the last section of the novel and they are not given much description or deliberation. The investigation, on the contrary, does seem more important and takes up a substantial part of the book. Therefore, Bolaño shifts the focus of his thriller from the crime to the quest.

Likewise, in *Los detectives salvajes*, the puzzle element of the detective story is saturated with a preoccupation for the literary, not only because the detectives are looking for a poet and her poetry, but because all the clues are framed in the context of literature. For instance, the ingenious deductions that the detectives are supposed to resolve in *Los detectives salvajes* are all puzzles on rhetoric and stylistics put by García Madero to the rest of the passengers of the Impala, such as 'les pregunté si sabían lo que era un gliconio (que es un verso de la métrica clásica que se puede definir como una tetrapodia logaédica cataléctica *in syllabam*)' [I asked them whether they knew what a glyconic was (it's a verse in classical meter that can be defined as a logaoedic tetrapody catalectic *in syllabam*)].[59] Thus,

58 Susanne Hartwig, 'Jugar al detective: el desafío de Roberto Bolaño', *Iberoamericana*, 28 (2007), 53–71 (56).

59 Bolaño, *Los detectives*, 558, 528.

the reader, who in a traditional thriller is invited to solve the case along with the detectives, is presented here with nothing but lampooned literary enigmas. A further example of 'literary detection' can be observed towards the end of their journey, when García Madero summarizes the clues they have come across: they found in a newspaper from 1928 that a relatively famous bullfighter has been seen in the company of 'Cesárea Tinaja'. García Madero explains how Belano and Lima respond to the description of the bullfighter in the article:

> a Belano y a Lima les recuerda la figura de un torero de Hemingway (autor al que desgraciadamente aún no he leído), el típico torero de Hemingway sin suerte y valiente y más bien triste, más bien mortalmente triste, dicen, aunque yo no me atrevería a decir tanto con tan poco en donde sostenerme, y además una cosa es Cesárea Tinajero y otra cosa es Cesárea Tinaja, algo que mis amigos pasan por alto achacándolo a una errata.
>
> [reminds Belano and Lima of a Hemingway bullfighter (Hemingway's an author I unfortunately haven't read), the typical brave and luckless Hemingway bullfighter, more sad than anything else, deathly sad, although I wouldn't dare say as much with so little to go on, and anyway Cesárea Tinajero is one thing and Cesárea Tinaja is another, which is something my friends refuse to admit, chalking it up to a misprint.][60]

Not only do Belano and Lima link the clues to the literary world via Hemingway's work (playing on the emblematic bullfighter of contemporary literature), but the hints themselves, as if they were a piece of literature, even contain typos. García Madero's objection about the image of the bullfighter and of the appearance of the name Cesárea Tinaja can be seen as the reluctance to accept a farfetched literary reading. Thus, although Bolaño's novel contains many elements of the thriller genre he also flouts many of the conventions. This is partly because the investigation is saturated with the literary, but also because of the movement of the narrative, which, instead of starting with the narration of the crime and moving backwards, places the deaths at the very end of the novel and gives them little, if any, consideration or analysis. In addition, the detectives are not salaried investigators or police officers or indeed, part of any legal apparatus; they

60 Bolaño, *Los detectives*, 571, 540–541.

are just young, marginal poets. Instead of having a mysterious crime, it is very clear who kills who and the investigation is not at all related to those crimes. Bolaño further denies the conventional expectations of thrillers and classic crime fictions in that there is no closure either at the end of the investigation or the end of the novel, there is no strong desire for justice on the part of the protagonists, nor is there a gradual increase of suspense as the plot progresses. Nevertheless, while Bolaño's novel does not adhere fully to any of the variants of the detective fiction genre, it does nevertheless contain certain of their elements as I mentioned above, namely, the two assassinations, an investigation, interrogations and 'detectives'.

As I have argued, the novel does not fall fully within any single genre rather, there is a simultaneous incursion into several. Returning to the extensive use of the verb 'perderse' seen in the novel we could interpret this as symptomatic of this type of incursion into and meandering between genres. In this way the novel explores an aesthetic of getting lost along with the disorientation of expectations that the entrance and exit into different genres can produce for the reader. *Los detectives salvajes* also blends the preoccupation for the investigation with that of the journey, offering an interesting intermingling of detective fiction and travel narratives. Ultimately the several genres in which the novel participates, travel and road narratives, adventure novels and crime fiction, tend to have in common the quest element, which is foregrounded in *Los detectives salvajes* at the time that a poetics of getting lost is underscored. This does not mean that the novel is not bound by genre formulas, but that both their use and their transgression have a purpose in the text. Through form, the aesthetic of getting lost and the meandering through generic expectations, the novel comes also to question Jameson's idea cited at the beginning of this piece, that third-world literature necessarily projects *a* 'political dimension in the form of national allegory'. The meandering of genres necessarily sways the economies of expectation generated by the text and invites plural responses, which could not possibly be distilled into the one 'national' reading. I have proposed several ways in which to read the journey in *Los detectives salvajes* by considering several features of the novel's form and theme: the speeding detachment form the Mexican landscape; the constant underscoring of being lost, travelling through the lines of poems written, smudged and lost and the frequent changes in genre. All of these readings of Bolaño's novel impede endorsing Jameson's proposal of Latin American literary works offering just one reading as national allegories. Literature in *Los detectives salvajes* is not wedded to the national, nor tied in with a continental project (within the novel, in fact, such readings are presented as highly inflated claims as to literature's scope and purpose) it is instead, sensitive to literature's own haphazardness.

CHAPTER 3

Juan José Saer's *El entenado* and the Failures of the Conquest

In an essay entitled 'Líneas del Quijote', Juan José Saer comments on a common thread in literary texts that emerges in Miguel de Cervantes's work and persists through to modern literature: the narrative of failure. According to Saer, failure overrides Cervantes's narration for, in opposition to the ends pursued by the hero in epic texts, Don Quixote's aims and objectives could never be fulfilled. Despite this dispiriting truth, Cervantes's protagonist sets out for and continues with his adventures. Saer aptly names Don Quixote's thrust 'moral del fracaso' [the ethics of failure], which for him defines 'toda la literatura occidental genuinamente representativa' [all genuinely representative Western literature].[1] In Saer's view, with variations and extensions, *the ethics of failure* is an undeniable central motif in the work of writers such as Flaubert, Kafka, Conrad, and Faulkner, amongst others.

With these reflections, Saer not only provides a telling connection in Western literature but throws some light on the interpretation of his own work. As Saer conceives it, the term *ethics* is not taken to designate a set of norms that restrict behaviour, but the very drive that impels a course of action, hence Don Quixote's continuous adventures. Thus, the *ethics of failure* refers to a code of conduct in which adventure (and, tacitly, travel) appears as an option that is both fundamental and bound to fail. At the same time, Saer's *ethics of failure* is connected to a broad body of literature, for, as seen in Cervantes's work, to embark on his adventures, Don Quixote draws his ideas from chivalric romances. But the text that narrates his adventures simultaneously invokes and deconstructs those romances,

1 Juan José Saer, 'Líneas del Quijote', in *La narración-objeto* (Buenos Aires: Seix Barral, 1999), 31–54 (47).

offering alternative fantasies, powerful yet pitiable, great imaginative feats, yet bound to fail.

The *ethics of failure* is likewise a well-suited term to introduce the analysis of *writing travel* in Saer's *El entenado.* As such, it provides an illuminating frame from which to read the journey and sojourn in America of the protagonist of Saer's novel. In a similar vein to that of Cervantes's text, *El entenado* also dramatizes its own reading of previous narratives, however these are not chivalric texts but those of the Spanish Conquest in Latin America. As I examine further below, the portrayal of imperial enterprises in *El entenado* does not foreground courage, the prowess of discovery and possession, but the very failure of those undertakings. In *El entenado*, thus, the *ethics of failure* relates to ways of writing the journey as well as reading preceding historical narratives.

Saer draws from limited historical information to write *El entenado*, but his novel also works with echoes and remainders of historical discourses in its aim to rescue from literature what was silenced by history. Saer explained in an interview that his novel stemmed from a paragraph in José Luis Busaniche's *Historia argentina*, which gives details of the ill-fated Spanish expedition of Juan Díaz de Solís to the River Plate region in 1521.[2] From that journey, in which almost all the crew were killed and then eaten by the Indians, only one member of those who disembarked survived and his name was Francisco Del Puerto. Del Puerto was a young cabin boy who was held captive by and dwelt with a tribe of anthropophagic Indians for ten years. After that he was found by Sebastián Caboto's expedition and it is uncertain whether he returned to Spain or not.[3] Saer commented that he had created and unfolded

2 Juan José Saer, 'Memoria del río', *Clarín*, 27 February (2000) <http://www.clarin.com/suplementos/cultura/2000/02/27/e-00501d.htm> [accessed 6 March 2009]; also in Juan José Saer, 'El valor del mito' in *La historia y la política en la ficción Argentina* (Santa Fé: Centro de Publicaciones Universidad Nacional del Litoral, 1995), 73–86; and José Luis Busaniche, *Historia argentina* (Buenos Aires: Solar-Hachette, 1973).

3 For a more detailed account on this see, Evelia Romano Thuesen, '*El entenado*: relación contemporánea de las memorias de Francisco Del Puerto', *Latin American Literary Review*, 23 (1995), 43–63.

his fiction drawing only on Busaniche's succinct documented information.[4] In this manner, *El entenado* weaves historical information into a fictional narrative situated in foundational times for Argentina. Though *El entenado* draws on this succinct historical account, it is a fictional piece, but one that also offers an assessment of the historical texts that emerged with the Conquest of America and its failures, for the story of captivity of the cabin boy in *El entenado* recalls the continual setbacks of colonialist ventures and the failure of their ideals. *El entenado* fails to constitute itself as a historical narrative, but in that very gesture, an ethics is at stake, which aims to recover the shadows of what history left in its wake. In Jacques Derrida's terms, *El entenado* emerges as a questioning of 'the spectres of history' for it attempts 'to ontologize remains, to make them present',[5] labouring its narration from a time 'out of joint': going back to what is not fully documented from the time of the Conquest and giving shape to a fiction that relies on the spectres and shadows of historical discourse. Saer's novel thus engages with the history of Colonial discourses of America (borrowing Fredric Jameson's words from another context), 'not as an event that *can* be narrated, but rather as an aporia around which the narrative must turn'.[6]

My aim in this chapter therefore is to examine the value of *El entenado* in appraising the *ethics of failure* in the textual production of the Spanish Conquest, while situating itself at a 'time out of joint'. Hence, a fundamental consideration will be the extent to which Saer's novel upholds or diverges from imperial paradigms. With this in mind, I will also discuss the consequences of Saer's specific representation of the Indians, and concepts related to the *writing of travel*: discovery, contact, conquest, and possession. First, in the following sections I explore the ways in which *El entenado* draws from, but also overturns, central tenets of the textual production that emerged from the Spanish Conquest.

4 Juan José Saer, 'Memoria del río', 74.

5 Jacques Derrida, *Specters of Marx*, trans. Peggy Kamuf (New York and London: Routledge, 1994), 9.

6 Fredric Jameson, *The Political Unconscious: Narrative as a Socially Symbolic Act* (London and New York: Routledge, 2002), 267 (emphasis is mine).

Reviewing the Spanish Conquest

El entenado is narrated by its nameless protagonist who, at the age of sixty, recounts the story of his sojourn in captivity with a group of Amerindians. He starts by relating his life as an illiterate boy who, driven by orphanhood and curiosity, is drawn to life in the Spanish ports. Having started with that picture, the narrative continues with the protagonist's days as cabin boy during his journey to the Americas. After the landfall, arrival and killing of most of the *entenado*'s companions, the novel is mainly shaped by the account of the main character's dwelling with the Colastiné Indians (who inhabited the regions surrounding the Paraná River, in Santa Fé, Argentina). Later in the novel he offers a more succinct and less detailed account of his return to Spain, his years in a monastery where a priest called Quesada teaches him to read and write, and his time as the writer, and subsequently travelling actor, of a play about his sojourn with the Colastiné.

Saer links the protagonist's writing of his journey to the Americas to different textual practices related to the Conquest's documents, such as the *cartas de relación*, *testimonios*, and *crónicas*.[7] It is important to take into consideration that all of the aforementioned productions were hybrid

7 Some examples of these texts are Columbus' *cartas* [letters], for example his *Carta-Relación del Almirante Don Cristóbal Colón a los Reyes Católicos sobre su tercer viaje* (1498), Alvar Núñez Cabeza de Vaca's *Relación* (later entitled and published as *Naufragios* (1542)), Bartolomé de las Casas' *Historia de las Indias* (1527), and Hernán Cortes' *Cartas de Relación* (1524). The *relación* [report] was a genre, a narrative format or form, where it was possible to accommodate a wide range of information that the conquistadors and Amerindians arranged and composed to recount the times of the Conquest. It was also regarded as a kind of legal statement and as such the text had to comply with the requirement of veracity. The *testimonio* was not a genre (strictly speaking), but it emerged as a form of writing more linked to authors with the aim of understanding cultures, social structures and religions, such as Bernardino de Sahagún, whose writings from *Historia general de las cosas de la Nueva España* (1569) were partially based on the declarations or 'testimonies' of native informants. The *crónicas* [chronicles] of the Indies were more akin to what is currently considered a historical account, since they were written more or less according to the conventions of modern Western history.

pieces of writing that modern readers would link not only to history, but also to discourses from other fields such as geography, anthropology, and natural sciences.[8] These texts presented the geography of territories seen as new to the Spaniards, together with their social, political and commercial organization. At the same time as they traced the dynamics of the processes of colonization, they showed the transformation of the conquerors, their perceptions of America and their views of the world. The significance of the connection between *El entenado* and the narratives of the Conquest partly also lies in the fact that the latter, as Stephanie Merrim states, 'framed the founding images and topics – of utopia, civilization and barbarism – which would resonate in European and Latin American writings, as well as influence colonialist policies, for centuries to come'.[9]

In his study of the narratives of travel in sixteenth-century America, Joan-Pau Rubiés observes that 'few sources reveal with the clarity of travel accounts the subtle tension between the projection of ideals [or myths] and the discovery of futility'.[10] In *El entenado* it is on the one hand the possibility of travel and, on the other, the results of the journey that mark that very tension. Thus, prior to the journey, travel and the New World are related to fictive images where newness and the marvellous prevail. For

8 For more on this see: Stephanie Merrim, 'The First Fifty Years of Hispanic New World Historiography the Caribbean, Mexico, and Central America', 58–100; and Kathleen Ross, 'Historians of the Conquest and Colonization of the New World 1550–1620', 101–142, both essays in *The Cambridge History of Latin American Literature*, eds Roberto González Echevarría and Enrique Pupo Walker, Vol. I (Cambridge: Cambridge University Press, 1996); also Naomi Lindstrom, *Early Spanish American Narrative* (Austin: University of Texas Press, 2004); Amy Fass Emery, *The Anthropological Imagination in Latin American Literature* (Columbia: University of Missouri Press, 1996); and Beatriz Pastor, *Discursos narrativos de la conquista: mitificación y emergencia* (Hanover: Ediciones del Norte, 1988) translated by Lydia Longhstreth Hunt as *The Armature of Conquest: Spanish Accounts of the Discovery of America, 1492–1589* (Stanford, CA: Stanford University Press, 1992).

9 Stephanie Merrim, 'The First Fifty Years of Hispanic New World Historiography', 59.

10 Joan-Pau Rubiés, 'Futility in the New World: Narratives of Travel in Sixteenth-Century America', in *Voyages and Visions: Towards a Cultural History of Travel*, eds Jás Elsner and Joan-Pau Rubiés (London: Reaktion Books, 1999), 74–100 (76).

instance, at the beginning of the novel, the prospect of the protagonist's and the crew's voyage creates in the imagination of the narrator a world of paradisiacal temptation. Thus, we read, 'ya los puertos no me bastaban: me vino hambre de alta mar […] en la orilla opuesta del océano y de la experiencia, la fruta es más sabrosa y más real' [by that time the ports no longer sufficed: I hungered for the open sea. far off on the other side of the ocean, on the farther shore of experience, the fruit is more succulent, more real].[11] The protagonist's anticipation for the journey and the promise of the New World in *El entenado* is comparable with early narratives of the Conquest (such as Columbus's letters) that, as Beatriz Pastor clearly points out, shaped a representation of America in which the imaginary and fictional prevailed over reality. As Pastor clarifies, this will to interpret and remodel the American reality in relation to myth responded to the need to adjust the vision of the New World to Spanish economic and ideological requirements.[12] Following the same line of analysis, Joan-Pau Rubiés observes that Columbus's first journal of discovery made of the blending of myth and history, a matrix that was to be followed in later narratives of discovery and conquest which combined 'the real Caribbean islands and mythical Cipangus, real gold and mythical el Dorados, real empires of the sun and mythical Amazons'.[13]

Saer retrieves a similar atmosphere, charged with fantasies, at the beginning of his novel when the protagonist reports the tales circulating in the ports. He mentions, for example, some of the images that were part of the

11 Juan José Saer, *El entenado* (Habana: Fondo Editorial Casa de las Américas, 2006), 26. All further references to this novel are taken from this edition. Subsequent page references in the text will refer first to the original and then to the translation taken from the English version *The Witness*, trans. Margaret Jull Costa (London: Serpent's Tail, 1990), 10.

12 Beatriz Pastor, *Discursos narrativos de la conquista*, 39. Giuseppe Mazzota's observation is akin to Pastor's view when he claims that with the arrival of Colombus in America, the 'new world' and the 'old world' formed a 'tangle of elements underneath the epoch-making phenomenon', in *New Worlds and the Italian Renaissance: Contributions to the History of European Intellectual Culture* (Leiden and Boston, MA: Brill, 2012), 9–24.

13 Joan-Pau Rubiés, 'Futility in the New World', 74.

inventory of the Americas that circulated at the time: 'ciudades pavimentadas de oro, del paraíso sobre la tierra, de monstruos marinos que surgían súbitos del agua y que los marineros confundían con islas' [cities paved with gold, of paradise on earth, of sea monsters that rose suddenly out of the water; sailors, mistaking these monsters' backs for islands].[14] The ultimate goal of the cabin boy's departure from Spain is the fulfilment of curiosity and the confirmation of what was imagined. Although the type of sea voyage that the protagonist recounts is one of monotony and boredom, those feelings act as a catalyst for the expectation of the apparition of the marvellous. As the protagonist states: 'A veces hubiésemos deseado, sin duda, la aparición de uno de esos monstruos marinos que llenaban la conversación en los puertos. Pero ningún monstruo apareció' [At times we almost longed for a sighting of one of those sea monsters so talked about in ports. But no monster appeared].[15] It is important to stress the fictional vein of these first images as the novel, with the course of its narration, seems to parallel the trajectory of the discourses of the Conquest which, as both Rubiés and Pastor explain, shifts from a highly fictionalized conception of the New World to one with a pervasive air of reality.

Accordingly, as the journey unfolds in Saer's novel we observe the entrance of a perspective which privileges what the protagonist calls 'the real' (which he sees as a tentative and provisional reality) and a progressive abandonment of a fantastic view. When the crew reaches the shores of America, the understanding of the New World takes on a strong empirical bent and here the novel proves imaginary travel to be an impossible enterprise, and the previous symbolic geography that was talked of in the ports is transformed into an experiential site. The arrival in the New World in *El entenado* heralds the beginning of the crisis of the imaginary and fabulous paradigms. This moment of change is marked in the narration as an encounter with origins, a progression from an abstract region towards a tangible place. And Saer uses the image of the river to express this transformation: 'El olor de esos ríos es sin par sobre esta tierra. Es un olor a origen, a formación húmeda y trabajosa, a crecimiento. Salir del mar

14 Saer, *El entenado*, 26, 10.

15 Saer, *El entenado*, 29, 13.

monótono y penetrar en ellos fue como bajar del limbo a la Tierra' [The smell of those rivers is like no other on earth. It is the smell of primeval beginnings, the dank smell of things painfully taking shape, caught in the very process of growing. Leaving behind the monotonous sea and travelling down those rivers was like descending from limbo on to earth].[16] This passage vividly captures the appearance of 'the real' and the falling away of the illusory. This departure from the illusive is further accentuated when the crew disembarks:

> Teníamos la ilusión de ir fundando ese espacio desconocido a medida que íbamos descubriéndolo, como si ante nosotros no hubiese otra cosa que un vacío inminente que nuestra presencia poblaba con un paisaje corpóreo, pero cuando lo dejábamos atrás, en ese estado de somnolencia alucinada que nos daba la monotonía del viaje, comprobábamos que el espacio del que nos creíamos fundadores había estado siempre ahí, y consentía de dejarse atravesar con indiferencia, sin mostrar señales de nuestro paso.
>
> [We nursed the illusion that by discovering this unknown land we were laying claim to it, as if before us there had been nothing but an immanent [*sic*] void which our presence peopled with corporal landscape. But when we left it, in that state of hallucinatory somnolence brought on by the monotony of the trip, we saw all too clearly that the space we considered ourselves the founders of had always in fact been there and had allowed our passage through it with indifference, retaining no trace of our passing.][17]

According to the conquerors' expectations, the place and the Indians would 'start to exist' when discovered by them. However, the passage also expresses the flaws inherent to imperialist claims: that is, in order to be plausible, the concept of discovery (associated in the passage with drowsiness and illusion) must deny the prior existence of the discovered place and its inhabitants.

Traditionally, the premise of the discovery, as Anthony Pagden observed, is that those who see a new land 'for the first time' feel able to claim the right of possession over it.[18] Hence, the process of discovery ends

16 Saer, *El entenado*, 39, 22.

17 Saer, *El entenado*, 38, 22.

18 Anthony Pagden, *European Encounters with the New World* (London: Yale University Press, 1994), 34.

with an act of possession. That notion contrasts (and will clash later in the novel) with the strength of an empirical nature which presents itself as totally indifferent to men. The last quoted passage describes a space unaltered by the crew's presence and view; it was already there and remains intact after they leave. The land of the Indians, therefore, eludes the expedition's discovery and possession. At this point we can see that the novel's itinerary, having begun in the realms of fancy, has led the travellers to a more experiential and empirical view of both the journey and its destination.

Simultaneous with the growing importance of an empirical view, the novel introduces the first hints of what constitutes the second type of Conquest narrative, what Pastor calls 'the discourse of failure', in which nature is commonly depicted as uncontrollable and hostile. This idea is invoked in *El entenado* when the Colastiné unreceptively meet the crew on the coast and kill them with bows and arrows, leaving the *entenado* as the only Spanish survivor. Although after a few years of dwelling with the Indians, the cabin boy articulates his changing view about Amerindian nature, indicating in turn a transition from a perception of its strangeness and ambiguity to clearer definitions of its attributes, his ongoing vulnerability persists. The land of the Colastiné is characterized thus as a site of hardship and shortage, and during his sojourn with the Indians, the protagonist experiences a permanent sense of the inviolability of nature. Necessity, suffering, hunger, and cold are chief elements of this part of the novel and, indeed, of the discourse of failure. The importance of this discourse is not only that it presents a counter narrative to valiant formulations of the Conquest, but that, as Pastor explains, it 'abre un espacio para la presentación objetiva y crítica de la nueva realidad' [creates a space where the objective and critical presentation of the new reality is possible] while inaugurating a discursive practice which formally establishes the first representations of America as detached from myth.[19] Pastor clarifies that the discourse of failure developed later on in the Conquest, but also co-existed with those accounts that drew from myth. A telling example of the discourse of failure is Hernán Cortés's *Quinta Carta de Relación* which shares

19 Beatiz Pastor, *Discursos narrativos de la conquista*, 191.

with the *entenado*'s account, as I explore further below, not only a more marked objective tone, but also the recognition that the American reality is a new one, and its differences cannot be accounted for in the writing.[20]

Between cultures

To an extent, *El entenado*'s account is also in keeping with what Mary Louise Pratt calls 'survival literature', in the sense that it is a first-person narration 'of shipwrecks, castaways', 'abandonment' and 'captivities'. For Pratt, survival literature creates a 'safe' context for 'staging the alternate, relativizing and taboo configurations of intercultural contact'.[21] These texts avoid being too transgressive for they 'presupposed the imperially correct outcome: the survivor survived and sought reintegration into the home society. The tale was always told from the viewpoint of the European who returned'.[22] Though also a tale of survival and captivity, the *entenado*'s text and status come precisely to question that imperially correct outcome noted by Pratt. With the journey, the *entenado* becomes a survivor, the one spared, yet he does not fit in with either the Colastiné nor his 'home' society. In this regard, Arcadio Díaz-Quiñones rightly points out that homecoming, or the reintegration of the *entenado* into his previous community, 'es imposible para el sujeto que ha circulado entre la "civilización" y la "barbarie"' [is impossible for the subject who has travelled between 'civilization' and 'barbarism'],[23] and therefore questions the possibility of seeing the *entenado*'s journey as one of completion and resolution. The *entenado*'s place of enunciation (in between the Amerindian tribe and

20 Beatriz Pastor, *Discursos narrativos de la conquista*, 197.

21 Mary Louise Pratt, *Imperial Eyes: Travel Writing and Transculturation* (New York: Routledge, 2008), 85.

22 Pratt, *Imperial Eyes*, 85.

23 Arcadio Díaz-Quiñones, '*El entenado*: Las palabras de la tribu', *Hispamérica*, vol. XXI, 63 (1992), 3–14 (12).

the Spanish) complicates the possibilities of having 'the viewpoint of the European who returned'; and the title of the novel, *El entenado*, is illuminating in this regard. As Rita De Grandis explains: 'in the Argentinean rural dialect, [entenado] connotes a negative value for a person we might call an orphan or a "bastard"'.[24] Amy Fass Emery's adds that the *entenado* 'is son and not son, being (ente-) and nothingness (nada/o)'.[25] And although the word *entenado* in the dictionary of the Real Academia Española is registered as stepson (*hijastro*), both De Grandis and Fass Emery recognize that the semantics of the title operate more within the realm of orphanhood or being an abandoned child. Moreover, both the protagonist's feelings of abandonment when the massacre of his companions occurs and the survivors on the ship leave, and the title of the book that the priest Padre Quesada wrote about him, *Relación de abandonado* [Report on the abandoned one], single the protagonist out while bringing to our attention that meaning of the title yet again. In so far as he maintains his status of *entenado/abandonado*, even after returning to Spain, the protagonist cannot, as Pratt points out commonly happened in 'survival literature', simply become one with or reintegrate into his home society on his return.

Instead of generating a 'safe context' for intercultural contact, the position occupied by the *entenado* after the journey is hence at an exclusionary boundary (geographical, cultural and ideological) between the Spanish and the Indians. When a new group of Spanish explorers arrive on Colastiné territory, the former cabin boy initially acts as an informer for the Spaniards, who rely on his testimony to explore the land and encounter the Indians. We learn in the novel that the Spanish captain cannot understand why the Indians hide in the face of the conquerors and the last encounter

24 Rita De Grandis, 'The First Colonial Encounter in '*El entenado*' by Juan José Saer: Paratextuality and History in Postmodern Fiction', *Latin American Literary Review*, 21 (1993), 30–38 (31).

25 Amy Fass Emery, 'Sa(l)vage Ethnography, The Cannibalistic Imagination in Juan José Saer's *El entenado*', in *The Anthropological Imagination in Latin American Literature* (Columbia: University of Missouri Press, 1996), 117. She has probably taken the idea from Raquel Fressard Linenberg's interview with Saer, *Río de la Plata*, 7 (1988), 155–159, in which Saer states that the meaning of the title is: 'El *Entenado*, ente y nada'.

between both groups is characterized as brief and violent and the result is the extermination of the Colastiné. The only image of this slaughter in the novel is offered through the account of the conquerors' journey (accompanied by the *entenado*) which is shown as traversing a river full of floating corpses that surround their craft.[26] In this way, the *entenado* comes to see the conquerors as lacking in heroism and common sense, undermining the valour of their occupation and conquest.

As both informer and captive, the protagonist is located between cultures and, consequently, is himself suspicious in the eyes of the conquerors. Recollecting his first meeting with the Spaniards he explains how much effort it takes him to speak Spanish and again the difficulties the captain and crew found in understanding him: 'Traté de hablar en mi lengua materna, pero comprobé que la había olvidado. Con gran esfuerzo, logré al fin proferir algunas palabras aisladas, formulándolas, por costumbre, con la sintaxis peculiar de los indios' [I tried to speak to them in my mother tongue but found I had forgotten it. At last, with great difficulty, I managed to utter a few isolated words, linking them out of habit in the idiosyncratic syntax of the Indians].[27] The protagonist's in-betweenness renders his language a hybrid that cannot be identified with any symbolic system or culture. Both the Spanish and the Indian environments are alien spheres to him. His condition, of not belonging to one culture or the other, renders him an abandoned man, an *entenado*, and therefore unsung by both societies.[28] After returning to Spain, the protagonist of Saer's novel continues his wanderings, first in the city and thereafter on his travels with a theatre company as an itinerant actor. His self-representation is

26 Brian Gollnick and other scholars have seen in this image an analogy with the corpses that appeared in the River Plate immediately after the military coup, in Brian Gollnick 'El color justo de la patria: agencias discursivas en *El entenado* de Juan José Saer', *Revista de crítica literaria latinoamericana*, 57 (2003), 107–124 (111).

27 Saer, *El entenado*, 111, 95.

28 In this regard, Luigi Patruno notes that the linguistic forgetfulness of the entenado thematizes a tension between 'lo propio y lo otro' [what is your own and the other's] that was also common in narratives of Conquest such as Álvar Nuñez Cabeza de Vaca's account, in Luigi Patruno, *Relatos de regreso: ensayos sobre la obra de Juan José Saer* (Rosario: Beatriz Viterbo, 2015), 58.

thus one of a traveller with no sense of, or longing for, a place. His story, as Iain Chambers puts it with regards to the story of migrancy, 'calls for a dwelling in language, in histories, in identities that are constantly subject to mutation. Always in transit, the promise of homecoming – completing the story, domesticating the detour – becomes an impossibility'.[29] Though the *entenado*'s voyage is in a way a return journey (he leaves and goes back to Spain) his placelessness and his perpetual status as an outsider blur his alleged homecoming.[30]

Returning to the civilization-barbarism dichotomy, pointed out by Díaz-Quiñones, we can see that it encapsulates two inextricable, yet incommensurable dimensions of the position of the protagonist on his return to Spain. In this, the protagonist-narrator reveals again to be 'out of joint', for that binary is key to illustrating the *entenado*'s dissenting views on how civilization and barbarism were regarded at the time of the Conquest. The novel describes how the condition of the Indians and their humanity was discussed by the ecclesiastical and political authorities and by other members of the church and the conquistadors: 'La condición misma de los indios era objeto de discusión. Para algunos no eran hombres; para otros, eran hombres pero no cristianos, y para muchos no eran hombres porque no eran cristianos' [The human condition of the Indians was the subject of much dispute. Some did not even consider them to be men; others thought that they might be men but that they were certainly not Christian men; and most thought that if they were not Christians they could not be men].[31] Those categorizations recall the Renaissance distinction between civilization and barbarism based upon religion, the identification of Christianity with civilization, and the consideration of the barbarian as the irrational being with mental and moral deficiencies who shares the behaviour of animals. Saer's protagonist, however, adopts a critical stance regarding that long-standing dichotomy. For him it is no longer clear that the Spaniards

29 Iain Chambers, *Migrancy, Culture, Identity* (London: Routledge, 1994), 5.

30 For more on this see, Arcadio Díaz-Quiñones, '*El entenado*: Las palabras de la tribu', 9; and María Cristina Pons, *Memorias del olvido: Del Paso, García Márquez, Saer y la novela histórica de fines del siglo XX* (Mexico: Siglo XXI, 1996), 225.

31 Saer, *El entenado*, 123, 108–109.

should be called humans since he thinks that 'no había más hombres sobre esta tierra que esos indios y [...] yo no había encontrado, aparte del padre Quesada, otra cosa que seres extraños y problemáticos a los cuales únicamente por costumbre o convención la palabra hombres podía aplicárseles' [the Indians were the only men on this earth, and [...], with the exception of Father Quesada I had met only strange problematic beings whom only custom or convention dignified with the name of 'man'].[32]

Although both the Indians' way of life and their anthropophagic habits were viewed in Europe as barbaric and, as a result, morally inferior, and thus the Indians were not seen to subscribe to the ideals of Western civilization (namely possessing technological, economic and artistic sophistication, being endowed with a writing system and institutions), it is significant here that the protagonist calls them the only men on earth. With this gesture, Saer clearly reverses the terms of the civilization-barbarism dichotomy and although this does not erode their antithetical relationship, it offers a turnaround of the long-standing binary. As such, the protagonist's views are forward-looking, for the stability of the meaning of the antinomy civilization-barbarism begins to be undermined, as Susan Sontag notes, with modern travel literature which 'starts when civilization becomes a critical as well as a self-evident notion – that is, when it is no longer so clear who is civilized and who is not'.[33] Hence the inversion of the opposition between savage and civilized signals the *entenado*'s alienation from his time, while also being a significant, deconstructive manoeuvre of Saer's novel.

The importance given in *El entenado* to specific informative and documentable facts also displaces the protagonist's account from the belief systems and literature of the Conquest. Throughout *El entenado* we find several references to the Colastiné peoples, written with an ethnographic

32 Saer, *El entenado*, 124, 109.

33 Susan Sontag, 'Questions of Travel', in *Where the Stress Falls* (London: Vintage, 2003), 274–284 (275). Susana Rotker has fittingly warned, however, that the civilization-barbarism dichotomy persisted as an exclusionary distinction in Argentine literature and culture well into the start of the twentieth century, in Susana Rotker, *Captive Women: Oblivion and Memory in Argentina*, trans. Jennifer French (Minneapolis and London: University of Minnesota Press, 2002), 11.

and curious tone, for instance their hygiene: 'En todos, el cuidado por la limpieza era excesivo, casi irritante' [All of them were excessively, almost irritatingly careful about cleanliness]; child-rearing methods: 'Sin ser duros ni autoritarios, el comportamiento con las criaturas era severo y, aunque no exento de consideración e incluso cariño, sentencioso y cortante' [Their treatment of their children was solemn, severe and curt, although neither harsh nor authoritarian, and certainly not without consideration or even affection];[34] treatment of captives: 'De tan serviciales, eran ostentosos y pesados' [They were obliging to the point of ostentation and tedium].[35] Their dress, behaviour, environment, climate, language (meanings, its capability for representing, and uses) and food are all commented upon in a similar way. The narrator also examines the Indians from a physical and moral point of view according to what he sees and experiences. However, the descriptions of the Amerindians do not include any details of the commercial possibilities for Spain and the narrator does not offer (nor indeed does he have) a religious justification or understanding of his journey. It is in respect of these two factors, commercial possibilities and religious justification, that *El entenado* can be differentiated from the literature of the Conquest, since those were dominant discursive paradigms in that body of work. In this sense, Saer delineates the world of the Spanish occupation in Latin America and constructs a counter-narrative pointing towards a different kind of colonial subject, whose concerns are neither commercial nor evangelical.

The protagonist's disposition to document certain facts and not others, together with him being an *entenado* of the imperial enterprise, displace him from the world, and ideology, of the Conquest. As the odd one out, he cannot belong to one society or the other. A further indication of his exclusion can be found in the text when the now adult cabin boy is reunited with the Spanish conquerors and he comments that they must see him as 'un extraño en ese lugar de pesadilla' [a stranger in that nightmarish place]. In other words, he is both strange to the Spanish and an outsider

34 Saer, *El entenado*, 86, 69.
35 Saer, *El entenado*, 87, 70.

in the Indians' space.[36] This idea of not belonging anywhere, either as a Spaniard or an Indian, is also emblematized in the novel's title, since a further interpretation of the word '*entenado*' is of someone who does not have a rightful place.

Writing travel

Despite the *entenado*'s not belonging to or integrating with the Indian domain, he becomes a special observer of the Colastiné and their land, and that is one of the reasons why he feels entitled to write a pseudo-ethnographic report on the tribe. The fact of being a witness, and of constructing an authorial voice which derives from personal experience and proximity to the situation described, is a device that is also present in the texts by chroniclers who wrote about the Spanish Empire in America. However, in the narratives of the Conquest, close observation went hand in hand with possession. In this regard, Stephen Greenblatt notes that, 'everything in the European dream of possession rests on witnessing, a witnessing understood as a form of significant and representative seeing'.[37] The narrators of the Conquest, being also possessors, acted as representatives of the Empire and, as Silvia Spitta argues, they 'confront[ed] the alterity of the New World from an identification with that imperial "I/eye"'.[38]

Although the narrator's voice in *El entenado* adopts the form of a personal testimony, his portrayal adheres to different needs to those of the Empire, one of them being the desire to justly represent the Indian other. Moreover, the protagonist is not only concerned with descriptions of the Indians but also with his role in their society. As narrator, he is devoted

36 Saer, *El entenado*, 111, 96.

37 Stephen Greenblatt, *Marvellous Possessions: The Wonder of the New World* (Oxford: Clarendon Press, 1991), 122.

38 Silvia Spitta, *Between Two Waters: Narratives of Transculturation in Latin America* (College Station: Texas A&M University Press, 2006), 32.

largely to his own recording of the past; therefore, his words take the fictionalized form of autobiography and self-definition. Seemingly contrary to the Colastiné's reality, the narrator's reality is also based on the preoccupation with the self (himself), his autobiography. His account, like any autobiography, is based on the conception of himself as an individual who belongs (or not) to a society but does not, however, form his voice from his experience within a collective set. Although some critics, such as Pons, have taken the protagonist's occasional use of 'nosotros' [we] to indicate some preference towards the notion of plural identity, I would contend that the protagonist never, in fact, assumes that plural or collective character.[39] The narrator expresses his concerns about putting together an accurate account of the Indians' past, but at the same time, the story of the Colastiné that he records – thanks to his memory – is his own life story. The *entenado* presents himself as the person who has travelled and survived, and then as the first (and the last) to write about the Colastiné. Travel and survival make it possible for him to preserve *his* Colastiné story. The narrator casts himself in the role of a traveller who 'solo en el mundo' [all alone in the world] tries to decipher his encounter with the Amerindians.[40] However, despite having lived with them for ten years, growing to know their way of thinking and living, he depicts them as a collective character. He is almost incapable of seeing individuals: there is not one example of special friendship or love affair during his whole sojourn with them. As such, it is difficult to agree with those critics who see an inclusive 'nosotros' [we] in the novel. Thus, although the *entenado* does not completely identify with the imperial 'I/eye', his story of cultural integration is also one of failure.

The Colastiné are interpreted and represented by the voice of an outsider, the *entenado*; and as mentioned earlier, the Amerindian voice itself is not articulated. The *entenado* paints a portrait of the Indians' society, and gives meanings and values to their customs and language which respond to his interpretations. The Indians form a collective almost deprived of individual features and they are depicted as incapable of seeing and understanding

39 Pons, *Memorias del olvido*, 237.

40 Saer, *El entenado*, 42, 26.

themselves objectively. Moreover, the protagonist contrasts their inability with the acuity of his own view:

> Lo exterior era su principal problema. No lograban, como hubiesen querido, verse desde afuera. Yo, en cambio, que había llegado del horizonte borroso, el primer recuerdo que tengo de ellos es justamente el de su exterioridad.
>
> [Their principal problem was the outer world. They could not, as they might have wished, see themselves from outside. On the other hand, the impression they made on me, a stranger from the far horizon, was precisely their externality.][41]

In his determination to represent others, the protagonist remains outside the Indians' universe. An attentive reader will observe that, though the protagonist-narrator aims to do justice to the Colastiné with his account, in his encounter with the tribe there is no real empathy with them and thus Saer's attempt to paint a 'collective character' also led to a homogenizing vision of the Indians. As Kimberle López notes, 'within *El entenado*, the ambivalence of the self-Other negotiations is apparent in many instances'.[42] The Indian other remains a single and collective other without individual differences.

Though there is no profound assessment of how the Indians could have seen the Spanish conquerors, we can tangentially gather the view of the Spanish which pertains to the Colastiné peoples from the novel's lengthy description of the Indians' language. This linguistic focus includes the possibilities of the Colastiné language to represent, conceptualize and name things, nature, and others, including their captives. The narrator explains that the Indians called him *Def-ghi*, which means, among other things, witness, because they expected that he would write about them. But by using the same word (*def-ghi*) for each of their captives, whether a Spaniard from halfway round the world or someone from a neighbouring tribe, we can see that, to the Colastiné, the Spaniards were just one among a possible

41 Saer, *El entenado*, 142, 128.

42 Kimberle López, 'Loving cannibalism: cannibalism and colonial desire in Juan José Saer's *El entenado*', in *Latin American Novels of the Conquest: Reinventing the New World* (Columbia: University of Missouri Press, 2002), 70.

range of others. But here we note that the narrator wrote as a result of the supposed hope placed in him by the Indians, and this is significant, for it clearly attenuates his association with colonialist discourses.

For Florencia Garramuño, the positioning of the narrator in *El entenado* also distances this text from Conquest narratives.[43] According to Garramuño, Saer makes clear that the narrator is writing a *version* of the Colastiné and that he intersects the 'ethnographic account' of the Indians with his own personal and partial report.[44] In doing so, the narrator questions the possibility of being fully objective regarding the Colastiné (or the past) and of offering a unique, valid version of what they were like and what happened. Notwithstanding the narrator's doubts towards the possibility of producing a mimetic rendering of the past and the Indians, Garramuño notes that Saer's writing makes use of realist devices that imply a will to recover the referential function of writing. In her own words, Saer's literary strategies: 'cuestionan la objetividad como posibilidad neutral pero al mismo tiempo recuperan cierta función referencial del relato y marcan la necesidad de seguir narrando a pesar de la crisis de la representación' [question objectivity as a neutral possibility but at the same time recover some referential function of the story and highlight the need to keep narrating in spite of the crisis of representation].[45] Garramuño's view reinforces my previous points concerning *El entenado*'s adoption of the *ethics of failure* in relation to representation. Here, the possibility of painting a final and accurate portrayal of the Colastiné is seen as an impossibility, any attempt to do so would end in failure, yet it is also 'an aporia around which the narrative must turn'.[46]

43 Florencia Garramuño, *Genealogías culturales: Argentina, Brasil y Uruguay en la novela contemporánea (1981–1991)* (Rosario, Argentina: Beatriz Viterbo Editora, 1997), 110.

44 In fact, as Garramuño remarks, the report that the 'padre Quesada' writes on the *entenado*'s experiences in America and the play written by the *entenado* himself (loosely based on his experiences in America), reinforce the point that the final narration that the *entenado* writes (and what we read) is just a version of his sojourn in America with the Colastiné.

45 Florencia Garramuño, *Genealogías*, 112.

46 Fredric Jameson, *The Political Unconscious*, 267.

Captivity

The narration's investment in conventions of captivity narratives is another feature that helps to confirm how the *entenado*'s journey is imbued with the *ethics of failure.* That tradition includes a variety of texts which commonly narrate the experiences of those who were captured and held captive by other peoples or groups. Fernando Operé describes captivity as a violent form of abduction frequent among North American Indian tribes before 1492.[47] From the time of their arrival in the New World, the Europeans also made use of captivity, for diverse reasons, but mainly as a form of obtaining an Indian labour force. Many Latin American tribes that were not used to taking captives started this practice at that time as a form of resistance, to harass the enemy, as a form of exchange or barter and to secure a supply of women for reproduction. To some extent, some captives were incorporated into Indian society and became both an important part of processes of transculturation and participants in the foundation of mixed communities. There were other captives, however, who never integrated and lived marginally as prisoners or slaves.

El entenado draws on some conventional features of captivity narratives, for instance, the beginning of the protagonist's experience with the Indians is portrayed as a rebirth to the world. As the *entenado* describes his first night with the Colastiné he states he cried and felt exceedingly lonely, as if he was witnessing his own birth. He goes on to explain that although he was already an orphan, he was born (again) under these special circumstances. Classically, the rebirth is described as the crossing of a dark threshold, and in Saer's novel its description is associated with the idea of a river with a maternal odour:

> Entenado y todo, yo nacía sin saberlo y como el niño que sale, ensangrentado y atónito, de esa noche oscura que es el vientre de su madre, no podía hacer otra cosa que echarme a llorar. Del otro lado de los árboles me fue llegando constante, el rumor

47 Fernando Operé, *Historias de la frontera: el cautiverio en la América hispánica* (Buenos Aires: Fondo de Cultura Económica, 2001).

> de las voces rápidas y chillonas y el olor matricial de ese río desmesurado, hasta que por fin me quedé dormido.
>
> [Though unaware at the time, I was a foundling and, like a dazed and bloody babe leaving the dark night of his mother's womb, all I could do was cry. From beyond the trees the sound of shrill, rapid voices and the uterine smell of that vast river kept drifting up to me, until at last I fell asleep.][48]

Captivity narratives rely on the account of severe and alienating experiences that the protagonists go through; in the case of the *entenado* he witnesses cannibal rituals in which his companions are eaten, and he suffers the hardships of a difficult life in the open air. With regards to the conventional motives for taking captives (barter or reproduction), the novel adds a new and original inflection on these. As I have mentioned previously, the protagonist becomes aware that he has been captured by the Colastiné in order that he should give an account of their story, and rescue them from oblivion through writing:

> De mí esperaban que duplicara, como el agua, la imagen que daban de sí mismos, que repitiera sus gestos y palabras, que los representara en su ausencia y que fuese capaz, cuando me devolvieran a mis semejantes, de hacer como el espía o el adelantado que, por haber sido testigo de algo que el resto de la tribu todavía no había visto, pudiese volver sobre sus pasos para contárselo en detalle a todos [...] querían que de su pasaje por este espejismo material quedase un testigo y un sobreviviente que fuese, ante el mundo, su narrador.
>
> [They wanted me to reflect like water the image they gave of themselves, to repeat their gestures and words, to represent them in their absence, and, when they returned me to my fellow creatures, they wanted me to be like the spy or scout who witnesses something that the rest of the tribe has not yet seen and retraces his steps and recounts it, meticulously [...] the Indians wanted there to be a witness to and a survivor of their passage through this material mirage; they wanted someone to tell their story to the world.][49]

The protagonist is made a captive in order to become the chronicler of the Indians' story in their 'pasaje por este espejismo material' [passage through

48 Saer, *El entenado*, 51, 35.

49 Saer, *El entenado*, 157, 144.

this material mirage]. Saer dramatizes the narrator's responsibility, being in the position to retrieve the 'shadows' of the history of an Amerindian tribe, tracing a path that originates with mimetic intentions ('to reflect like water'), falling, inevitably, into mediated representation: a re-tracing and a re-counting of the Colastiné story. It is the *entenado*'s exterior viewpoint, his lack of integration with the Indians' community and finally their complete extermination that impel and allow the *entenado* to embark on his own *writing of travel*, as a tracking down of the vestiges of a vanished reality. The word mirage in the passage above, together with the provisional quality of the reality set down by the *entenado* are fundamental, for they offset the weightiness of the 'ethnographic' descriptions of the tribe. If a positive outcome of the journey is the accomplishment of a direct observation of the Amerindian tribe, one of its negative outcomes is that what is observed and written is after all a 'mirage'.

There are other senses in which *El entenado* as a captivity narrative actualizes the *ethics of failure*. In Latin America and North America captivity narratives generally constituted a type of discourse that was connected to the frontiers or to what Pratt calls the 'contact zones'. This latter term is better suited to define the encounters and the clash of cultures experienced by captives because, as Pratt explains, while the notion of colonial frontier is 'grounded within a European expansionist perspective [...] "contact zone" shifts the center of gravity' and 'foregrounds the interactive, improvisational dimensions of imperial encounters so easily ignored or suppressed by accounts of conquest and domination told from the invader's perspective'.[50] Indeed, there were 'contact zones' wherever the Indians put up resistance, first to the European, and afterwards to the Creole expansion.[51] In consonance with Pratt, Operé notes that captives were the 'víctimas silenciosas de las problemáticas relaciones de la frontera, los ignorados protagonistas del encuentro en el Nuevo Mundo. Sus nombres y sus historias pasaron

50 Mary Louise Pratt, *Imperial Eyes*, 8.

51 For instance in Argentina it was thought that a 'frontier' separated the territories of the Creole and the Indians until the final extermination of the latter with the military campaign in 1879 against the Mapuche, Tehuelches and Ranqueles called 'La campaña del Desierto'.

inadvertidos en el gran trasiego que representó el choque de culturas' [silent victims of the problematic relationships of the frontier, the ignored protagonists of the encounter in the New World. Their names and their stories passed unnoticed in the hustle that the clash of cultures represented].[52] Operé also reminds us that captivity was a ubiquitous phenomenon in the frontiers of southern Chile and the Argentine Pampas.

Captives' stories started to appear in the chronicles during the sixteenth century and there were still present in some travel accounts in the eighteenth and the nineteenth centuries, such as Lucio V. Mansilla's *Excursión a los indios Ranqueles* (1870) and Alfred Ebélot's *Relatos de la frontera* (1968), which mentioned the existence of captives among the Indian peoples. Nevertheless, even though captives were taken for extended periods of time, there are few written examples of non-fictional stories of captivity (written by captives, ex-captives or others) in Latin America. The reason for that lack is, in Operé's view, that Spain was more interested in accounts of discoveries or conquest rather than in successes of the Indian resistance. In light of that, Operé considers that all the writings about the Conquest and the New World were expected to be narratives of an expansive civilization, and not narratives about its failures. With that idea in mind, any history or account of white captivity in Latin America can be seen as a testimony to the failures of the Spanish Empire. Viewed this way, all such narratives of captivity are allied very closely to Saer's *ethics of failure* and to Pastor's characterization (explored above) of a whole corpus of Conquest's texts as discourses of failure.

While the importance of most captive stories prior to the independence of Latin American countries was underplayed by the chroniclers, captivity narratives later became, mostly during the nineteenth century, an important theme within the field of Argentine fiction. In this sense, Saer's literature follows a common strand in Argentine literature in its aim to rescue what historical discourses disparaged.[53] Thus, it is likely that

52 Fernando Operé, *Historias*, 9.

53 As Operé explains, captivity narratives were normally sequels in chronicles and the narration was written in the third person, as happens in Gonzalo Fernández de Oviedo's *Historia general y natural de las Indias* III (Madrid: Biblioteca de Autores

the main effect Saer intended, when choosing to write a fictional story of captivity supported by a voice which narrates in the first person, was to retrieve, through fiction, a story linked to those captives who normally passed unnoticed in colonial times. In other words, the text could be aiming to shed light on histories that were commonly subdued, but that were to become, centuries later, key in Argentine literature.

I began this chapter with reflections on the *ethics of failure*, and how it acts on the relationship that Saer's novel has with the writing of history, and I would like to return now, as I conclude, to those ideas. As I demonstrated above, the *ethics of failure* is a common thread in narratives of travel that register the abandonment of triumphalist and fantastic paradigms and contribute to tracing the losses that imperial enterprises signified for the New and the Old Worlds. As such, the *ethics of failure* emerges from a clash between ideals and the encountered American realities. Rubiés fittingly notes that this clash was 'not only the tragedy of wanton cruelty and unintended genocide which we encounter if we focus our gaze on the American Indians but also the tragedy of the European identities – religious, political and social – projected upon the new continent'.[54] In this

---

Españoles, 1959), 28. Accounts in the first person like Nuñez Cabeza de Vaca's or Francisco Nuñez Pineda de Bascuñán's *Cautiverio feliz* were an exception. The centrality of fictional captivity narratives has been highlighted by Jean Franco when claiming that white-women captives' stories in Argentina were, 'one of the pathetic figures of contemporary folk-lore'. Esteban Echeverría's *La cautiva* (1837), one of the first pieces of fiction exploring the story of a woman captive, is probably the best-known example of this type of narrative, along with some of the characters in emblematic books such as José Hernández's *Martín Fierro* (1872) and Ascasubi's *Santos Vega o los mellizos de la flor* (1872). This narrative strand has been enlarged and continues to the present day in contemporary Argentine literature. See for instance Jorge Luis Borges' 'Historia del guerrero y la cautiva', in *El aleph*, 1st edition 1949 (Buenos Aires: Emecé, 1957), 47–52; César Aira's *Ema la cautiva* (Buenos Aires: Editorial de Belgrano, 1981); Leopoldo Brizuela's *El placer de la cautiva* (Buenos Aires: Temas, 2001), Graciela Montaldo's and Claudia Gilman's *Preciosas cautivas* (Buenos Aires: Alfaguara, 1993), Guillermo Saccomanno's *La lengua del malón* (Buenos Aires: Planeta, 2003), Sylvia Iparaguirre's *La tierra del fuego* (Madrid: Suma de letras, 2001) and Martin Kohan's *Los cautivos* (Buenos Aires: Sudamericana, 2000).

54 Rubiés, 'Futility in the New World', 76.

sense, the travel narratives explored here: survival literature, captivity narratives and more generally the discourses of the failure of the Conquest, highlight failure, while building upon the tension between the need and desire to travel and the encounter with the New World.

At the same time, *the ethics of failure* impels the *entenado* to *write travel*, to record the history of the Colastiné. Here, again, the story is not one of triumph but one of failure. The past events, and the Colastiné, are portrayed as an 'espejismo material' [material mirage]; they are, therefore, inaccessible and non-representational. However, the writer aims to approach them by fictionalization and textualization. In this regard, in his 1994 article entitled 'El valor del mito', Saer asserts the impossibility of writing historical novels: for him the concept of the 'historical' alongside 'novel' is a fallacy or, as he puts it, 'una imposibilidad epistemológica' [an epistemological impossibility].[55] In effect, he posits an interpretation of fiction not as more-or-less historical, but as working with the unfeasibility of narrating history within fiction. *El entenado* writes travel, rendering the story of the Colastiné as 'mirages', rather than registering a different historical version, or set, of events. Writing the history of a success of the Conquest, affirming the possibility of grasping the reality of the Colastiné, or sustaining the historical novel's epistemological possibility, would deny *the ethics of failure* at stake in *El entenado*.

55 Juan José Saer, 'El valor del mito', in *La historia y la política en la ficción argentina* (Santa Fé, Argentina: Centro de publicaciones universidad nacional del litoral, 1995), 73–86 (74).

CHAPTER 4

Hidden in a Desert Cave: Reading and Disclosure in Roberto Bolaño's *2666*

> APOCALYPSE means simply Revelation, though there is nothing simple about this one, since men have puzzled their brains for nearly two thousand years to find out what exactly is revealed in all its orgy of mystification.[1]
>
> — D. H. LAWRENCE

Roberto Bolaño's *2666* (2004) is an unfinished text but the longest, most elaborate and probably the best of his novels.[2] Central to this text are a particular sense of time, which is provisional and scarce, and a striking construction of spaces, observable chiefly in the Sonoran Desert and a border city called Santa Teresa. Both time and space bear an intrinsic relationship to the idea of an ending, and both relate to a tradition of reading endorsed by the novel which is, as a great part of the critical scholarship on *2666* remarks, apocalyptic. As I argue further below, Bolaño's borrowings from the Apocalypse are not based on a consideration of it as a repository of religious truths, but on its imaginative literature of covert sense. Crucial to the novel's apocalyptic vision is its ability to reflect on the drastic consequences of advanced capitalism in peripheral territories, political corruption, and issues of violence now characteristic of the border zone where a great part of the novel is based, and which include 'illegal transnational

1 D. H. Lawrence, *Apocalypse* (London: Penguin, 1980), 3.

2 Roberto Bolaño, *2666* (Barcelona: Anagrama, 2004). All further references to this novel are taken from this edition. Subsequent page references in the text will refer first to the original and then to the translation taken from the English version *2666*, trans. Natasha Wimmer (London: Picador, 2009).

capitalism based on the trade in narcotics and other outlawed activities'.[3] Although in common parlance 'apocalyptic' has been and is used just as a synonym of a catastrophic or calamitous finale, an aspect that has not been acknowledged by most critics who qualify *2666* as apocalyptic is that this adjective can carry with it as much a 'sense of an ending' as a feeling of a new beginning or renewal. Drawing on this last aspect of the apocalyptic, it is my contention that while Bolaño's proposal in *2666* aims to interrogate the status quo, it also offers oppositional readings from where we can derive imaginative and hopeful inspiration. Therefore, a cornerstone of this chapter is the special connection that exists between *2666* and apocalyptic tropes which not only involve the idea of a catastrophic finale, but also denote that the text we are to examine contains meanings yet to be unveiled. With this perspective in mind, my focus in *Writing Travel* shifts here to the examination of the writing of spaces, mainly that of the desert and its connections to the apocalyptic.

It is symptomatic that in a review of Sergio González Rodríguez's book entitled *Huesos en el desierto*, Bolaño argues that there are only two existing traditions in Latin America: 'la tradición aventurera' [the adventurous tradition] and 'la tradición apocalíptica' [the apocalyptic tradition] probably because 'son las únicas que nos acercan al abismo que nos rodea' [they are the only ones which get us closer to the abyss that surrounds us].[4] Though I would not go so far as identifying only two traditions of writing in Latin America, I am interested in highlighting the importance given by Bolaño to the apocalyptic tradition through which he posits that one can understand cultural productions. In relation to the connection between Bolaño's novel and the apocalyptic tradition, Ignacio Echevarría comments that: 'En una de sus abundantes notas relativas a *2666* Bolaño señala la existencia en la obra de un "centro oculto" que se escondería debajo de lo que cabe considerar, por así decirlo, su "centro físico"' [In one of his numerous notes for *2666*, Bolaño points out in his novel the existence of a 'hidden center', which would be concealed beneath what can be considered (in a

3 John Kraniauskas, 'Border Issues', *Travesia* 3:1–2 (1994), 5–13.

4 Roberto Bolaño, 'Sergio González Rodríguez bajo el huracán', in *Entre paréntesis* (Barcelona: Anagrama, 2004), 214–216 (215).

manner of speaking) the novel's 'physical center'].[5] Thus Bolaño presents his novel as containing a 'hidden centre'; and invites the reader to engage in an interpretation exercise in which (as with the apocalyptic tradition) to read becomes very much to decipher, or to attempt to unveil the 'centro físico' of the novel.

Other studies linking *2666* to the apocalyptic tradition include Edmundo Paz Soldán's 'Introducción, Roberto Bolaño: literatura y Apocalipsis'. In Paz Soldán's view, a connection to the apocalyptic tradition can be found not only in *2666* but also in much of Bolaño's earlier work and he suggests that it has its roots in the history of Latin America, or rather, in the experiences of the dictatorships and their expression in Latin American literature that Bolaño explores in texts such as *Estrella distante* (1996) and *Nocturno de Chile* (2000).[6] As Paz Soldán puts it, 'En el escritor chileno [...] no hay otra opción que dar cuenta del horror y del mal, y hacerlo de la manera excesiva que se merece: el imaginario apocalíptico es el único que le hace justicia a la América Latina de los años setenta' [In the Chilean writer [...] there is no other option but to give an account of the horror and evil, and to do it in the excessive way that it deserves: the apocalyptic imaginary is the only one that does justice to the Latin America of the seventies].[7] He reinforces the link between Bolaño's work and the Apocalypse based on *2666*, a book which is defined by Paz Soldán as 'la aventura y el Apocalipsis, diseminados a lo largo y ancho del planeta' [the adventure and the Apocalypse, disseminated across the length and breadth of the planet].[8] However, Paz Soldán's observations, although valid, leave the specific apocalyptic features of Bolaño's literature unidentified.

Cathy Fourez agrees that the apocalyptic is a main feature in *2666*. She believes that this characteristic emerges from the narrative of the extermination of women in the novel. Fourez connects the numbers 666, which

5 Ignacio Echevarría, 'Nota a la primera edición', in Roberto Bolaño's *2666* (Barcelona: Anagrama, 2004), 1123.

6 Edmundo Paz Soldán, 'Introducción, Roberto Bolaño: literatura y Apocalipsis', in *Bolaño salvaje* (Barcelona: Candaya, 2008), 11–30.

7 Paz Soldán, 'Introducción', 13.

8 Paz Soldán, 'Introducción', 19.

appear in the novel's title, to the landfills described in the novel where the bodies of assassinated women are dumped, qualifying this space as a 'territorio apocalíptico' [apocalyptic territory].[9] In addition, Fourez observes that, when the critics are in Santa Teresa, the sky of Bolaño's novel, 'se asemeja al del águila del capítulo VIII del Apocalipsis de San Juan, que augura en voz alta el infortunio de los habitantes de la tierra' [is similar to that of the eagle of the eighth chapter of the Apocalypse of Saint John, which foretells aloud the misfortune of the earth's inhabitants].[10] Therefore, in that scholar's view, the sky in that part of the novel reflects the atrocity of the women's deaths. Other scholars have established further links between this novel and the Apocalypse, for instance, Ángeles Donoso Macaya suggests that there is a relationship between the main city of *2666*, Santa Teresa, and Gabriel García Márquez's Macondo. For Donoso Macaya, Macondo represents 'El origen mítico de Latinoamérica' [the mythical origin of Latin America] and 'Santa Teresa puede ser apreciada como su posible final apocalíptico' [Santa Teresa could be considered as its possible apocalyptic end].[11] While I do not necessarily share Paz Soldán's conclusion about the exclusive connection between the Apocalypse and Latin American literature of the 1970s, and while unlike Fourez I do not consider the apocalyptic features to be limited to the relationship with the crimes (although this is a fundamental connection), the links established between the use of apocalyptic images and Bolaño's work in those scholars' studies have been revealing and helpful for my study.

Taking a different stance from the work reviewed above, Patricia Poblete suggests a connection between Bolaño's personal circumstances and ideas of Apocalypse explored in *2666*. According to this scholar:

> En el Apocalipsis coinciden la muerte del individuo y el fin del mundo, y, por consiguiente, los objetivos personales y comunales quedan inextricablemente unidos:

9 Cathy Fourez, '*2666* de Roberto Bolaño: los basureros de Santa Teresa, territorios del tiempo del fin', in *Los imaginarios apocalípticos en la literatura hispanoamericana contemporánea* (Bern: Peter Lang, 2010), 231–244 (233).

10 Fourez, '*2666* de Roberto Bolaño', 238.

11 Ángeles Donoso Macaya, 'Estética, política y el posible territorio de la ficción en *2666* de Roberto Bolaño', *Revista hispánica moderna*, vol. 62, 2 (2009), 125–142 (139).

> el apocaliptista no separa la historia de la biografía. Lo mismo sucede con Bolaño: *2666* marca el fin de su escritura y también el fin de su vida.
>
> [In the Apocalypse the death of the individual and the end of the world coincide, and therefore, personal and communal objectives end up being inextricably joined: the 'apocaliptista' does not separate history from biography. The same occurs with Bolaño: *2666* marks both the end of his writing and of his life.][12]

Poblete asserts that a main part of the meaning of Bolaño's novel can be explained by extra-textual context, that is, Bolaño's sickness and acknowledgement of his proximity to death. She therefore defines the novel as a 'suerte de Apocalipsis' [a kind of Apocalypse] in that *2666* is a stage for death – 'las de las mujeres en Santa Teresa, la del propio Autor' [that of the women in Santa Teresa and of the author].[13] Like Paz Soldán, Poblete contends that these images and tropes of death are not new to Bolaño's work; indeed, they are a recurring theme which can be identified in his previously published novels. Poblete's analysis of the existing link between death (in the novel and in the author's real life) and apocalyptic literary features is thought-provoking, however I do not agree with her remark that *2666* 'cancela la posibilidad de su experiencia histórica' [cancels the possibility of historical experience], since to speak about the Apocalypse and Bolaño's biography as fictions of the end interlaced with the novel is necessarily to conceive them as a historical experience. Poblete goes on to say that in the novel 'se liquida, además, cualquier perspectiva de sentido y de trascendencia' [any possibility of making sense, and of transcendence, is annihilated],[14] notwithstanding that to explain *2666*'s style and aesthetics through Bolaño's proximity to death when writing the novel and its derivative apocalyptic motifs is, indeed, to give meaning and sense to the novel. As Frank Kermode notes in a different context, 'apocalyptic types [...] decadence and renovation, progress and catastrophe – are fed by history and underlie our ways of making sense of the world from where we

12 Patricia Poblete Alday, *Bolaño: otra vuelta de tuerca* (Santiago de Chile: Universidad academia de humanismo cristiano, 2010), 139.

13 Poblete, *Bolaño: otra vuelta de tuerca*, 10.

14 Poblete, *Bolaño: otra vuelta de tuerca*, 138.

stand, in the middest'.[15] My thesis thus coincides with Kermode's point and differs from Poblete in that I consider that *2666* proposes a type of reading that has to do with disclosing meanings and restoring sense, independently of how unstable or not these meanings are. I believe that, through the apocalyptic, the novel does not adhere to the argument that there is no meaning, on the contrary, it puts forward a fictional practice that has the potential to re-envision spaces and events in ways that generate probing and meaningful questions (and at times answers) in its readers. In order to better locate my commentary on the specific ways in which the novel develops writing practices related to the apocalyptic tradition, I offer below a brief summary of the novel structure and plot.

2666

Although unfinished at the time of Bolaño's death in 2003, the 1,119-page novel, *2666* (Spanish edition) is an ambitious and accomplished text. The work is divided into five 'parts' or sections which, while they at times converge, Bolaño designed and wished to be published separately (mainly for financial reasons that would benefit his heirs). John Kraniauskas argues in this regard that Bolaño's novel 'is an inorganic work' which 'does not quite make a whole, and whose unity is given paradoxically in narrative proliferation and dispersal'.[16] The five parts of the novel are entwined through characters which might form 'the background in one part' but 'the foreground in another'.[17] For Kraniauskas those connections 'emphasize the autonomy of parts – what is central to one part is not to another – but, as the text refocuses, endow it with a certain three-dimensionality or

15 Frank Kermode, *The Sense of an Ending: Studies in the History of Fiction* (Oxford: Oxford University Press, 2000), 29.

16 John Kraniauskas, 'A Monument to the Unknown Worker: Roberto Bolaño's *2666*', in *Radical Philosophy*, 200 (2016), 37–46.

17 Kraniauskas, 'A Monument', 37.

depth of field'.[18] This depth of field is also achieved, I believe, through the combination of multiple temporalities in the novel. Although in overall terms *2666* breaks from chronology, as parts of the last section take place historically before events in the first section, each individual section follows a linear structure. If each part seems to reinforce the idea of chronological development, the multifocal and non-linear structure of the novel as a whole can be seen, in Lois Parkinson Zamora's and Walter Benjamin's terms, as revealing 'the falsity of the aesthetic unity to which the "classical" work of art pretends'.[19] To be sure, the temporal structures of the novel (discussed in more detail later in this chapter), both in its entirety and its sections, are likewise suggestive of Bolaño's conceptualizations of time in *2666* which combine the sense of an ending and visions of doom with liberation and openness.

The first section of the novel, 'The Part about the Critics', focuses on the lives of four European scholars, at the same time that it offers a humorous exposé of the academic world. Here four academics, Norton, Pelletier, Espinoza, and Morini, interact in their passion and search for the mysterious, missing writer known by his *nom de plume* of Benno von Archimboldi. They regularly travel to conferences where they present their enquiries and appreciation of this writer's work. These journeys are normally accompanied by extended walks, conversations and meals in different cities and ultimately an increasingly confusing tangle of their sexual and emotional ties. Not much is known about the life and person of Archimboldi, just that he is tall, has blue eyes and has not been seen for many years by anyone who knows of his identity except his publisher. After hearing a somewhat dubious tale about the author's possible whereabouts three of the four scholars set off to follow his trail which takes them to the Mexican border town of Santa Teresa in the Sonoran Desert. During this part of their quest they meet another academic, Amalfitano, who teaches at a local University and becomes the central character of the second part of the novel.

18 Kraniauskas, 'A Monument', 37.

19 In Lois Parkinson Zamora's *The Usable Past: The Imagination of History in Recent Fiction of the Americas* (Cambridge: Cambridge University Press, 1997), 166.

The second section, 'The Part about Amalfitano', begins with this character musing about what it was that drove him to Santa Teresa. During long internal monologues, Amalfitano takes the reader through flashbacks of his previous life in Europe, his memories of his ex-wife who was troubled by bizarre thoughts and experiences (and left him to raise their daughter alone). He also reflects on the nature of time and space, tells of unsettling events such as the appearance of a geometry book of whose provenance he is ignorant and how he has a compulsion to write lists of philosophers that he puts together almost involuntarily. The narrative takes on an unnerving atmosphere with the arrival on the scene of the son of the Dean of Sonora University, who seems to be stalking Amalfitano. Amalfitano's daughter, Rosa, starts dating an American sports journalist who goes to Santa Teresa to cover a boxing match. This journalist is initially called Quincy Williams and later Fate and he is the main protagonist of the third section. Fate's part also starts with this character's reflections, which trigger reminiscences about the death of his mother and his work at a New York magazine called *Amanecer negro* [*Black Dawn*], of which the main readership is the black community. His coverage of the boxing match leads Fate to be unwittingly caught up in the crime wave currently engulfing the city of Santa Teresa.

In the fourth section, called 'The Part about the Crimes', the course of the narration devolves into a detailed description of crimes taking place in Santa Teresa – including the multiple killings of women. Besides explicit reports of the crimes, this section also tells, amongst others, the stories of some of the investigating police officers, how one of the police detectives and the director of a mental hospital have a romantic liaison, the tale of a clairvoyant called Florita Almada and of the deeds of a man who desecrates churches. Finally, in the last section of the novel, 'The Part about Archimboldi', the reader is given a vivid presentation of this enigmatic, fictional writer's biography. Here, the novel recounts how Archimboldi grew up fascinated with marine life and how he left school early and worked as a servant before being drafted into the German army in 1939. The narration then moves on to his days as a soldier fighting for Germany in the Second World War, where we get a glimpse of the battles he was involved in and detailed accounts of the back scenes of war. At one point he needs to recover from a wound and, holed up in an abandoned house in the Russian village of Kostekino, Archimboldi discovers the hiding place and notes of Boris Ansky, a Jewish-Soviet writer, which were to have a great influence on his life and thoughts. After the war, Archimboldi is reunited with his girlfriend and starts his career as a writer. Finally, towards the end of his life, Archimboldi is drawn towards Mexico. Here the disparate strands of the novel begin to come together and Archimboldi finds himself in the very same desert that provides the core setting for the rest of the novel's stories and their cast of characters.

Title and epigraph

Kraniauskas argues that 'apart from its apocalyptical and biblical contents, the novel's title, *2666*, seems to demand that we look upon the present of its writing – our present – as if a past historical epoch', thus Bolaño's novel 'can be read both as a historical novel of the future and as a work of science fiction'.[20] However, I believe that it is within, and not apart from its apocalyptic contents that the title invites us to see the novel as a critique of the order of things (as if they were past) but also explore possibilities from which the future can be derived. Borrowing Fredric Jameson's words on science fiction, the title *2666* seems to be 'a form which now registers some nascent sense of the future, and does so in the space on which a sense of the past had once been inscribed'.[21] To be sure, the title *2666* immediately calls for the novel to be interpreted and analysed since, at first sight, while the numbers of the title might be a future date, they cloud the expectancy of the reader as they bring no clear and final meanings of the novel into the open; rather, they conceal them. One interpretation, that I follow in my analysis below, posits this number as a date in the future, as Kraniauskas pointed out. Another take on it could follow from fragmenting the number

20 Kraniauskas, 'A Monument', 39.

21 Fredric Jameson, 'Progress Versus Utopia: Or, Can We Imagine the Future?', *Science Fiction Studies*, Vol. 9 N. 2 (1982), 147–158, 150.

2666 and re-reading it as two times 666, forming, with the last three numbers, what is stated in the Apocalypse as the number of the beast or the Antichrist. The relevant verse of the book of Revelation reads: 'Wisdom is needed here; one who understands can calculate the number of the beast, for it is a number that stands for a person. His number is six hundred and sixty-six' (Rev. 13:18).[22] But this line has been interpreted in many ways and with very different results as to what it might mean. In this respect William Barclay notes that 'it is almost certainly true that more ingenuity has been expended on this verse than on any other verse in Scripture'.[23]

One of the first and most popular interpretations of that verse was obtained by assigning letters to each of the numbers that appear in it according to the traditional Hebrew coding system known as gematria or gimatria. In this system, each consonant (and then words and phrases) is assigned a numerical value which are then added together. According to the methods of gematria, the number 666 might be de-coded as the Hebraic name of Nero Caesar, NRWNQSR. Here the appropriate solutions would be N 50 R 200 W 6 N 50 Q 100 S 60 R 200, which add up to 666.[24] Eventually, the number 666 also came to designate any entity or person linked with the Antichrist (there have been associations of 666 with the Pope, Adolph Hitler, John Knox, Napoleon and many others).

But what does this number represent in Bolaño's novel? A possible explanation emerges from the fact that the number 666 is in itself the double of 333 (the number of the Trinity and therefore the number of God). This doubling has been related to the devil's pride and desire to be the double of or more than the Almighty. Following this logic, we could say that the title of the novel (*2–666*) is a doubling of the already doubled number 666. Whatever the case, the question remains as to why Bolaño chose that number or date for the title of his novel. If we agree with Theodor Adorno's assertion that a title acts as the 'microcosm of the work' then,

22 Taken from the New Revised Standard Version of the Bible, Anglicized Edition.

23 William Barclay, *The New Daily Study Bible: The Revelation of John*, Volume 2 (Louisville, KY and London: Westminster John Knox Press), 113

24 For more on this see Luis H. Rivas', *¿Qué dice el Apocalipsis?* (Buenos Aires: Pan y Trabajo, 1987), 13–14.

even if we do not know what the numbers 2666 mean, it is clear that they indicate the existence of a code that is there to be deciphered or revealed,[25] and thus, from the outset of the novel Bolaño seems to be positioning the reader to take on a 'revelatory reading'. The number *2666* is not repeated or alluded to at all in the body of Bolaño's novel, so there is no definite and final clarification of what the number stands for, hence the invitation to the reader to decipher its meaning persists throughout (and probably after reading) the novel.[26]

In the notes to the first edition of the novel Echevarría states that another of Bolaño's novels, *Amuleto* (1999), could offer an interesting insight into what the title might mean. Here, once again, the suggestion is to interpret the title *2666* as a date. The protagonist of *Amuleto* states that an avenue called Guerrero 'se parece [...] a un cementerio [...] de 2666 [...] olvidado debajo de un párpado muerto o nonato, las acuosidades desapasionadas de un ojo que por querer olvidar algo ha terminado por olvidarlo todo' [is more like a cemetery [...] in the year 2666 [...] forgotten [...] under the eyelid of a corpse or an unborn child, bathed in the dispassionate fluids of an eye that tried so hard to forget one particular thing that it ended up forgetting everything else].[27] Although the extract from *Amuleto* does not provide a final answer as to what the title means,

25 Theodor W. Adorno, *Notes to Literature*, Volume Two. Trans. Shierry Weber Nicholsen. Ed. Rolf Tiedemann (New York: Columbia University Press, 1992), 4.

26 Although there is a hint to the numbers in the title in the Part of Archimboldi when one of the members of the military tells the story of a mathematician who 'durante los últimos veinte años de su vida se había dedicado a buscar "unos números misteriosos", que están ocultos en alguna parte del vasto paisaje visible para el hombre, pero que no son visibles [...] una matemática alternativa camuflada [...] a la espera de que un hombre sea capaz de verla y descifrarla. El único problema era que para descifrarla había que verla y que para verla había que descifrarla' [spent the last twenty years of his life devoted to the search for some 'mysterious numbers' hidden in a part of the vast landscape visible to man, though the numbers themselves were invisible [...] mathematics camouflaged [...] just waiting for the man capable of seeing it and deciphering it. The only problem was that to decipher it one had to see it and to see it one had to decipher it], 858–867.

27 Bolaño, *2666*, 1124, 897.

it is however important in that it relates the number 2666 to death, forgetfulness and a date in the future. These three elements grant the title a sense of predicting the future and also link it, through the image of the forgotten cemetery, to the femicides narrated in 'The Part about the Crimes' in *2666*.

I would like to dwell here on how both title and crimes are ciphered in the novel as containing or covering up a secret, a feature about which the reader is warned: 'nadie presta atención a estos asesinatos, pero en ellos se esconde el secreto del mundo' [No one pays attention to these killings, but the secret of the world is hidden in them].[28] Emilio Sauri points out in this regard that the 'secret' in *2666* can be explained by the fact that 'Bolaño's fictional bordertown is one of the many black holes of global capitalism into which entire populations disappear'.[29] Patricia Espinosa takes a different stance in her article entitled 'Secreto y simulacro en *2666* de Roberto Bolaño'.[30] Here she also delves into 'el problema del secreto' [the problem of the secret] in *2666* and *Los detectives salvajes*. Although her reflections are not specifically about the novel's title, *2666*, it is useful to mention her study here as Espinosa sees, in Bolaño's play with mystery, a crucial aspect of those two novels, namely that the secret in Bolaño's texts is affected by the writer's manipulation of the reader's 'ansia desesperada de referente' [desperate longing for a reference].[31] In Espinosa's view the mystery in the novels is not to be resolved, 'no quiere ser descubierto' [it does not want to be discovered] for 'el secreto se mantiene siempre como indeterminación. Porque, para Bolaño, lo indeterminado es la vida y la determinación es la muerte' [the secret is always maintained as indeterminate. Since for Bolaño, life is indeterminate and what is determined means death].[32]

28 Bolaño, *2666*, 239.
29 Emilio Sauri, 'Autonomy after Autonomy, or The Novel Beyond Nation: Roberto Bolaño's *2666*', *Canadian Review of Comparative Literature*, 42 (2015), 396–409, 403.
30 Patricia Espinosa, 'Secreto y simulacro en *2666* de Roberto Bolaño', *Estudios filológicos*, 41 (2006), 71–79 (79).
31 Espinosa, 'Secreto y simulacro', 79.
32 Espinosa, 'Secreto y simulacro', 79.

While I find Espinosa's commentary on the primal importance of the secret or mystery in Bolaño's novels insightful (for the reasons I explained above about the type of reading that the text invites one to do), I do not share her view that the secret in *2666* is not to be unveiled, or to use her words, discovered. In consonance with Sauri's point above, Kraniauskas's article counteracts Espinosa's view on the complete lack of determinacy, for he argues that *2666* 'reveals', in fact, a 'public secret': 'that the world of narco-accumulation in which the murder of women takes place is also a dense and complex narco-*culture*; that is, in almost the classical sense of the term "a whole way of everyday life" (and death) – with its rituals, symbols and institutions, including its own laws'.[33] Like Kraniauskas and Sauri, I would agree that the way in which the novel presents how 'the killing of women is one aspect of an entire culture'[34] cannot possibly be sustained over the will to remain completely indeterminate. As Sauri notes 'what *2666* elicits, in this sense, is comprehension' for 'while it cannot quite tell us how each of these women come to meet such horrific ends in the black holes of contemporary capitalism, it does suggest that the novel [...] can tell us what our relationship to those victims cannot be'.[35] Though the reader never discovers the true identities of the perpetrators of the crimes, we cannot deny that they are caught up in the conditions of rampant capitalism (both legal and illegal) in the border town and that we are, as readers, involved in a reading that seeks to invest meaning to those circumstances. Thus, the secrets, to which the title and the crimes refer in the novel, do not drive the reader to the terrain of utter vacillation, but to a reading (or readings) imbued with the will to understand the 'complex and dense narco-culture' of the border from which Bolaño's novel draws.

If the title of *2666* is critical, the epigraph also illustrates the novel's relationship with apocalyptical readings and the crimes. It reads: '*Un oasis de horror en medio de un desierto de aburrimiento*' [An oasis of horror in a desert of boredom]. This line of verse is taken from a poem by Charles Baudelaire entitled 'Le voyage' and it expresses an overturning of expected

33 Kraniauskas, 'A Monument', 43.

34 Jean Franco, *Cruel Modernity* (Durham, NC: Duke University Press, 2013), 239.

35 Emilio Sauri, 'Autonomy after Autonomy', 406.

attributes: an oasis is normally a place of calm which contains water and vegetation and it is in stark contrast to the arid desert. In this verse, however, the oasis is one of horror and differentiated from the vastness of the desert, which itself is characterized by boredom. Baudelaire's poem appears in *Les fleurs du mal*, a book which deals with the subject of evil, and in this particular poem, evil is also related to travel.[36] In 'Le voyage' a traveller, moved by the longing for the unknown, chooses travel as an escape from tedium. The desire to discover something new (as with any other desire) is insatiable, and disappointment and boredom are unavoidable. Nevertheless, the quest continues in the poem and produces further experiences, such as the encounter with horror. As a whole, in Baudelaire's poem, to travel brings about an uneasy revelation: the world is monotonous, small; and it also shows us a part of our own image, our common use of horror.

We can gain some insight into the epigraph from Bolaño himself. In his essay entitled 'Literatura + enfermedad = enfermedad', in which he comments on Baudelaire's 'Le voyage' he states that there is probably no better word to define or diagnose the sickness of modern man than boredom. In addition, Bolaño asserts that 'Hoy, todo parece indicar que solo existen oasis de horror o que la deriva de todo oasis es hacia el horror' [Today, everything seems to indicate that only oases of horror exist or that which derives from every oasis is horror].[37] Although it appears that Bolaño seems to be positing that any escape from 'the modern sickness' leads to horror, I

36 Charles Baudelarie, *Les fleurs du mal* (Paris: Gallimard, 2004).

37 Roberto Bolaño, *El gaucho insufrible* (Barcelona: Anagrama, 2003). Bolaño ends the explanation of Baudelaire's poem with the following assertion: 'En un oasis uno puede beber, comer, curarse las heridas, descansar, pero si el oasis es de horror, si solo existen oasis de horror, el viajero podrá confirmar, esta vez de forma fehaciente, que la carne es triste, que llega un día en que todos los libros están leídos y que viajar es un espejismo. Hoy, todo parece indicar que solo existen oasis de horror o que la deriva de todo oasis es hacia el horror' [At an oasis one can drink, eat, heal the wounds, rest, but if the oasis is one of horror, if only oases of horror exist, the traveller will be able to confirm, this time in an indisputable way, that the flesh is sad, that there is a day in which all the books are read and that travelling is a mirage. Today, everything seems to indicate that only oases of horror exist or that the direction of everything is towards horror] (152) (my translation).

think that this is an overstatement more related to Baudelaire's work and the image of *poète maudit* that Bolaño is trying to fit into, rather than an authentic aesthetic and political proposal. Terry Eagleton characterizes the attitude seen here, in Bolaño's reading of 'Le voyage', as part of 'the jaded sensibility of postmodern culture', which 'no longer finds much shock value in sexuality, so it turns instead to evil, or at least to what it guilelessly imagines to be evil'.[38] In his use of Baudelaire's verse, Bolaño also seems to be seeking this shock effect referred to by Eagleton. As I will explore further, however, by developing complex settings in *2666* – principally the desert and Santa Teresa – through which Bolaño proposes ways of reading that suggest an opening in the system – ways out of the 'oasis of horror' – he is offering meanings that exceed the profound pessimism expressed in the idea that horror is the only alternative that exists today.

Another question that emerges from the epigraph is: what is the significace of this oasis of horror located in the middle of a desert of boredom? One possible interpretation of the oasis, despite its being full of horror, is that it is also the place where meaning resides. This reading stems from the fact that the oasis is being differentiated from a desert of *boredom*, where, as Lars Svendsen observes, boredom has long been linked to 'loss of meaning'.[39] However, this is not the full story, for, as Svendsen also argues, 'boredom always contains a critical element, because it expresses the idea that either a given situation or existence as a whole is deeply unsatisfying'.[40] Hence, boredom 'can be understood as a discomfort which communicates that the need for meaning is not being satisfied'.[41] In this way, we could see the desert of boredom as an invitation to consider that very 'need for meaning'.

As we read on, the motifs of the desert, horror and boredom are reiterated and their relationships to each other explored. For instance, in the last section of *2666*, Archimboldi's sister, Lotte, receives news about her son who is held in prison in Mexico as the main suspect of the

38 Eagleton, *On Evil*, 122.

39 Lars Svendsen, *A Philosophy of Boredom* (London: Reaktion Books, 2005), 18.

40 Svendsen, *A Philosophy*, 22.

41 Svendsen, *A Philosophy*, 30.

crimes. Lotte is very concerned with the news and has a dream in which she sees her brother walking through the desert, where the bodies of women were dumped. She cannot comprehend the meanings contained in that desert and characterizes it as 'tierra incomprensible y hostil' [an unfathomable and hostile land].[42] In the dream, Archimboldi replies to her as if communicating a secret: '–No – le decía Archimboldi, pero se lo decía como al oído – esta tierra es sobre todo aburrida, aburrida, aburrida' ['No', said Archimboldi, and he seemed to whisper this in her ear, 'it's just boring, boring, boring'].[43] Then there are the words of the clairvoyant in 'The Part of the Crimes', who claims on TV to have seen 'en el rostro del aburrimiento cosas horribles que prefería no decir' [terrible things in the face of boredom, things she would rather not recall].[44] Although the clairvoyant does not fully disclose her visions, the horrors contained in the face of boredom are, of course, the femicides committed in the desert. This horror restates that of the novel's epigraph and raises concerns with the identity, or identities, of the perpetrator(s) of the femicides. It is, however, not only the identity of the perpetrator(s) of the femicides which need to be unveiled; as Kraniauskas points out, the bodies of women also signal 'one of the many stories left untold in the novel', in the sense that their lives pre-mortem and their identities are uncounted for.[45] Thus, though without a complete, fixed, sense, we could also see the epigraph and the extracts above through Patricia Meyer Spacks' observation that: 'boredom presents itself as a trivial emotion that can trivialize the world'.[46] Thus the oasis of horror – and the femicides – essentially different from the desert of boredom, are therefore matters that should not be trivialized.

42 Bolaño, *2666*, 1101, 879.
43 Bolaño, *2666*, 1101, 879.
44 Bolaño, *2666*, 542, 433.
45 Kraniauskas, 'A Monument', 45.
46 Patricia Meyer Spacks, *Boredom: The Literary History of a State of Mind* (Chicago and London: The University of Chicago Press, 1995), 13.

Writing the desert and Santa Teresa

The Sonoran Desert is the material source on which Bolaño draws to construct the surroundings of Santa Teresa. This borderland, which forms part of the United States-Mexico frontier, is a region of 120,000 square miles which includes south-western Arizona in the United States, a great part of Baja California and half of the state of Sonora. It was delimited after conflicts and clashes including a war between Mexico and the United States (1846–1848) which ended with the Guadalupe-Hidalgo Treaty in which Mexico lost all of their territorial possessions north of the Río Bravo (about half of their total land surface). This border is also known as a frontier or barrier for undocumented migration from Central America and Mexico into the United States, a fact that has generated controversy and hostilities between the two nations. The borderland has been characterized in very different and sometimes contradictory ways. As Claire Lindsay has pointed out, the Mexico–US border 'has been variously conceptualised as a historical wound, a den of sexual licentiousness, an Edenic Paradise, and, in the decade of the 1990s as a privileged theoretical site'.[47] Apart from the more artistic and theoretical approaches to this territory, the Sonoran Desert has also been closely looked at more recently by the international press for its relationship with drug trafficking and the cartels. As mentioned above, it is also the site where multiple unsolved crimes take place; it is, therefore, a vastly contested area.

Bolaño acknowledges these realities in his construction of the fictional Santa Teresa. In addition, he depicts the city's urban sprawl – due in part to the constant stream of migrants attracted to work in the maquiladoras (US-owned manufacturing plants).[48] For Sharae Deckard, the setting of the novel in a border town like Santa Teresa exposes 'the systemic violence

47 Claire Lindsay, *Contemporary Travel Writing of Latin America* (New York: Routledge, 2010), 92.

48 In Melissa W. Wright's, 'Feminine Villains, Masculine Heroes, and the Reproduction of Ciudad Juárez', *Social Text*, 69 (2001), 93–113 (94), we read that 'almost 50,000 people migrate to Ciudad Juárez annually, many of them seeking employment as unskilled maquiladora laborers'.

of millennial capitalism', since this space 'makes profoundly tangible the tragic asymmetries and experiential contradictions of the periphery's relation to the core'.[49] In the novel, Santa Teresa is related to an 'impoverished' mobility through its inhabitants, who are largely described as poor refugees, transitory beings living in a provisional and horrible settlement. For example, when the scholars enter the city we read that 'la ciudad les pareció un enorme campamento de gitanos o de refugiados dispuestos a ponerse en marcha a la más mínima señal' [the city looked to them like an enormous camp of gypsies or refugees ready to pick up and move at the slightest prompting].[50] Santa Teresa is a constantly expanding city which sprawls into the desert, a quality that the scholars also notice: 'tuvieron la certeza de que la ciudad crecía a cada segundo' [They were convinced the city was growing by the second].[51] We also read that Amalfitano sees the city as a 'populosa ciudad que desafiaba al desierto, en la frontera de Sonora y Arizona' [populous city that stood in defiance of the desert on the border of Sonora and Arizona].[52] The crowding of the city is thus seen as challenging the desert rather than smoothly melding into it.

In this way, Santa Teresa is a place of contrasting and conflicting qualities. On the one hand, in *2666* one of the city's determining characteristics is to be 'horrible' (Liz Norton is one of several characters who repeatedly describe the city as such), and when tracing the missing writer, Archimboldi, the scholars wonder what could have brought anybody to such a city. On the other hand, the city is one of the most economically attractive in Mexico since, as one character of the novel notes 'Aquí casi todas las mujeres tienen trabajo. Un trabajo mal pagado y explotado [...] pero trabajo al fin y al cabo, lo que para muchas mujeres [...] es una bendición' [Here almost all women have work. Badly paid and exploitative work, [...] but work after all, which is a blessing for so many women].[53]

49 Sharae Deckard, 'Peripheral Realism, Millennial Capitalism, and Roberto Bolaño's *2666*', *Modern Language Quarterly*, Vol. 3, 3 (2012), 351–372 (353).

50 Bolaño, *2666*, 149, 111.

51 Bolaño, *2666*, 171, 129.

52 Bolaño, *2666*, 244, 189.

53 Bolaño, *2666*, 710, 568.

Santa Teresa is a constantly growing urban area that 'challenges the desert', making their mutual boundaries uncertain. It is also located at the border between Mexico and the US, and is the centre of migratory waves of poor people, mostly women, that come to work in the maquiladoras. The desert and the city then, should be seen as sharing fundamental characteristics: both are places of crime, both have blurred boundaries so that it is difficult to distinguish the limits between the city and the desert, yet both at the same time have a boundary: they constitute the border between Mexico and the United States. Thus, the dualistic thinking that divides space into urban (Santa Teresa) versus natural (the desert) is questioned here with the overlapping of one space into the other. Likewise, despite the clear political boundary to the north of the border, the scholars' observation of Santa Teresa as expanding thwarts the spatial closure and separation necessary for thinking of this area as a complete case apart. In what follows I examine how the desert, in all of its expressions, registers the narrative impulse of the novel in which significant meanings appear enshrouded, buried or hidden.

A space hiding meaning

One of the ways in which Bolaño's novel establishes an association between the desert and the apocalyptic is through the importance of light in the description of that terrain. In the Apocalypse, as Barclay has pointed out, 'John depicts God in "the flashing of gem-like colours" since it was "a familiar theme in the Bible to see God in terms of light. The Pastoral Epistles describe God as dwelling in 'unapproachable light'" (1 Timothy 6:16), and in the Psalms God is described as wrapping himself in light as if in a garment (Psalm 104:2)'.[54] This Book also states that light emanating from the Lamb of God will allow for the deciphering of the secret of the sealed Book (the Book of God) and that after the end of times everything is light, night no longer exists. The ancient fight between light and darkness finishes with the triumph of the former.

54 Barclay, *The New Daily Study Bible*, Vol. 1, 165.

Light, its effects and the images it makes are a key part of how the desert is described in *2666*. I will now analyse this facet of the novel, beginning with the following example which gives the reader their first introduction to the Sonoran Desert:

> Al salir del aeropuerto los tres percibieron la luminosidad del estado de Sonora. Era como si la luz se sumergiera en el océano Pacífico produciendo una enorme curvatura en el espacio. Daba hambre desplazarse bajo aquella luz, aunque también, pensó Norton, y tal vez de forma más perentoria, daba ganas de aguantar el hambre hasta el final.
>
> [As they left the airport, the three of them noticed how bright it was in Sonora. It was as if the light were buried in the Pacific Ocean, producing an enormous curvature of space. It made a person hungry to travel in that light, although also, and maybe more insistently, thought Norton, it made you want to bear your hunger until the end.][55]

In this passage, the main attribute of Sonora is 'la luminosidad' [brightness] and one of its associated terms is the resulting 'enorme curvatura en el espacio' [enormous curvature in space]. The notion of bending light to bury it in water (the Pacific Ocean) brings to mind the way in which mirages are formed. Thus, we have a suggestion that the first impression that the scholars get of the Sonoran Desert is an optical illusion. The scholars move, or travel, through the desert under a special light, 'aquella luz' [that light], which is powerful (and indeed is referred to as such by many characters in the novel) and it is associated with hunger. In the extract above it is stated that to travel in that light generates 'hambre' [hunger] but, as the character Norton thinks, it also 'daba ganas de aguantar el hambre hasta el final' [made you want to bear your hunger until the end]. This association between travel, the desert and resistance to hunger 'until the end' is reminiscent of both the biblical exodus, that is, the forty years of wandering in the desert, and Christ's forty days of fasting in the desert, on Mount Sinai, understood by the Christian exegesis as a parallel to the exodus. In both cases hunger is suffered, a hunger that was inflicted by God to teach a lesson, and both of these episodes of hunger were preparatory times of hardship to understand that obedience to God. The suffered hunger was less important when seen as a passage to get

55 Bolaño, *2666*, 147, 110–111.

to the Promised Land or to obtain freedom from sin. Of course, a capability to endure hunger is key to attaining passage in both cases.

From the first sight of the desert, we see the concept of endurance as embedded in the novel. It is also especially relevant to the Apocalypse, being highlighted as a key religious virtue: in Revelation 1:9, John defines himself as 'your brother and companion in tribulation, in the kingdom, and in that steadfast endurance which life in Christ alone can give'. By definition, endurance entails the expectation of the realization of a historical end or plan. Therefore, the significance and possibility of endurance is determined by ascribing teleological import and order to events. In *2666*, Bolaño offers other (at times playful) allusions to endurance connected to the desert. For example, in the second part of the novel we find Amalfitano thinking that the deserted landscape was 'apto solo para jóvenes o para viejos imbéciles o viejos insensibles o viejos malvados dispuestos a infligir o infligirse una tarea imposible hasta el último aliento' [best suited to the young or the old, imbecilic or insensitive or evil and old who meant to impose impossible tasks on themselves and others until they breathed their last].[56] This geography thus imposes demands on people that, though put humorously, recall Bolaño's own experience of writing the novel, namely, endurance through illness. These images of endurance, despite their being portrayed as 'imbecilic or insensitive', highlight purpose, which is also an emphasis in Bolaño's own narrative.

On many occasions the tone of the descriptions of the desert in the novel starts as objective, but typically develops into a fantastic or oneiric mode. By way of example, we see that in a three-way dialogue between the journalist Fate, a cook and a waiter, where they speak about the sunsets in the desert; everything seems normal until the end of the conversation when Fate sees that 'el desierto empezó a desvanecerse' [the desert began to disappear].[57] This disappearance of the desert relies on the interplay between cognition and estrangement, generating a distanced but also puzzling view of that space. It also makes us realise that what started as a realistic description of a particular landscape ends up, as in the scholar's view of

56 Bolaño, *2666*, 263, 205.

57 Bolaño, *2666*, 342, 269.

the desert as a mirage quoted above, being something illusory and closer to a dream. This connection between the desert and dreams is something that the novel constantly reinforces and reminds us of the ways of reading proposed by the novel, for dreams (since Freud at least) are to be disclosed and interpreted. As Paul Ricoeur has observed, 'the dream and its analogues are thus set within a region of language that presents itself as the locus of complex significations where another meaning is both given and hidden in an immediate meaning'.[58] Moreover, although what is disclosed or revealed is different in dreams (unconscious content) compared to religious texts (a 'divine' message), both, as Ricoeur notes, share a type of hermeneutics where 'to interpret is to understand a double meaning'.[59] It is therefore particularly significant that in *2666* dreams are constantly associated with the desert, since this relationship puts in motion the idea that this space should also be interpreted, analysed, and unfolded.

Another instance in which the desert and dreams are linked is found in the third part of the novel when Fate is about to cross the desert: 'Pensó que en algún momento, entre una y otra, había soñado con ese paisaje, no tan oscuro, no tan desértico, pero ciertamente similar' [At some point in between childhood and adolescence, he thought, he had dreamed of this landscape or one like it, less dark, less desertlike].[60] In that section of the novel we observe how Fate engages with the desert, going from a non-definitive perception of the site to the certainty of having dreamed about it. In his attention to the place, he links his desire to escape from the darkness of the desert with old memories from childhood and adolescence and in a very birth-like dream, he recollects and equates growing up with him emerging from the obscurity of that primal deserted territory. There are other passages in *2666* which reinforce the oneiric representation of the desert space. One such passage occurs in the first part of the novel and deserves particular attention in the way that it adds to the meanings already attributed to the desert. Here, the scholars have recently arrived

58 Paul Ricoeur, *Freud and Philosophy: An Essay on Interpretation* (New Haven, CT and London: Yale University Press, 1970), 7.

59 Ricoeur, *Freud and Philosophy*, 8.

60 Bolaño, *2666*, 343, 270.

in Santa Teresa, Mexico and we learn that the Spanish scholar, Espinoza, has a dream featuring the desert:

> Espinoza soñó con el cuadro del desierto. En el sueño Espinoza se erguía hasta quedar sentado en la cama y desde allí, como si viera la tele en una pantalla de más de un metro y medio por un metro y medio, podía contemplar el desierto estático y luminoso, de un amarillo solar que hacía daño en los ojos, y a la figuras montadas a caballo, cuyos movimientos, los de los jinetes y los de los caballos, eran apenas perceptibles, como si habitaran en un mundo diferente del nuestro, en donde la velocidad era distinta, una velocidad que para Espinoza era lentitud, aunque él sabía que gracias a esa lentitud, quienquiera que fuera el observador del cuadro no se volvía loco. Y luego estaban las voces. Espinoza las escuchó. Voces apenas audibles, al principio sólo fonemas, cortos gemidos lanzados como meteoritos sobre el desierto y sobre el espacio armado de la habitación del hotel y del sueño. Algunas palabras sueltas sí que fue capaz de reconocerlas. Rapidez, premura, velocidad, ligereza. Las palabras se abrían paso a través del aire enrarecido del cuadro como raíces víricas en medio de carne muerta.
>
> [Espinoza dreamed about the painting of the desert. In the dream Espinoza sat up in bed, and from there, as if watching TV on a screen more than five feet square, he could see the still, bright desert, such a solar yellow it hurt his eyes, and the figures on horseback, whose movements – the movements of horses and riders – were barely perceptible, as if they were living in a world different from ours, where speed was different, a kind of speed that looked to Espinoza like slowness, although he knew it was only the slowness that kept whoever watched the painting from losing his mind. And then there were the voices. Espinoza listened to them. Barely audible voices, at first only syllables, brief moans shooting like meteorites over the desert and the framed space of the hotel room and the dream. He recognized a few stray words. *Quickness, urgency, speed, agility.* The words tunnelled through the rarefied air of the room like virulent roots through dead flesh.][61]

The representation of the desert is constituted by Espinoza in his dream of himself contemplating the painting (as if it was a screen) located in the hotel room where he is sleeping. This juxtaposing of images and spaces, I believe, articulates a break with a conception of mimesis, and allows for the conception of a multi-perspectival space and time. The symbolism of the horsemen is an obvious link to the Apocalypse – as is the description of a space so intensely luminous that 'que hacía daño en los ojos' [it hurt your

61 Bolaño, *2666*, 154, 114–115 (emphasis in the translation).

eyes]. The 'jinetes' [horsemen] are conceived in the dream as if they were living in a different and slower realm, however, voices that announce the speediness of time connect the remoteness of their ambit to the present, as the voices are shot into the desert, making their way through it, into the hotel room and the painting. Thus, as where Fate has a conversation in which the desert seems to disappear (quoted above), the images and sounds in the dream of the desert open up the limitations of mimetic representation by way of their un-realistic ways of depicting time and space.

The desert is portrayed in more realistic terms in its role as a site for committing the femicides and dumping the bodies, a connection for which no easy or clear explanation is offered. For instance, Fate asks in a conversation with a character from Sonora (Chucho Flores):

> – ¿Y cómo las matan? – preguntó Fate.
> – Eso no está nada claro. Desaparecen. Se evaporan en el aire, visto y no visto. Y al cabo de un tiempo aparecen sus cuerpos en el desierto.
>
> ['And how are they killed?' asked Fate.
> 'Nobody's sure. They disappear. They vanish into thin air, here one minute, gone the next. And after a while their bodies turn up in the desert'.][62]

The importance of the term 'disappeared' in describing the bodies must be stressed here. This description crops up on other occasions throughout the novel, for instance, later in 'The Part of the Crimes' a politician who has been searching for a friend declares: 'y una mañana o una noche mi amiga se desvanece en el aire' [And one morning or night my friend vanishes into thin air].[63] Likewise, the identities of many victims of the rapes and murders remain unknown, forgotten and 'disappeared' from the registers: 'que quedaron fuera de la lista o que jamás nadie las encontró, enterradas en fosas comunes en el desierto o esparcidas sus cenizas en medio de la noche, cuando ni el que siembra sabe en dónde, en qué lugar se encuentra' [those who didn't make it onto the list or were never found, who were buried in unmarked graves in the desert or whose ashes were scattered in the middle

62 Bolaño, *2666*, 363, 287.
63 Bolaño, *2666*, 786, 629.

of the night, when not even the person scattering them knew where he was, what place he had come to].[64] The novel makes it clear that, if the identity of each disappeared body is to be reinstated, it must first be unearthed. Here, we observe how the narrative pulse of the novel, in which reading entails disclosing, is manifested in a physical and more realist manner to evaluate some of the consequences of the femicides.

The theme of the disappeared, also explored by Bolaño in novels such as *Nocturno de Chile* and *Estrella distante*, connects with the experience of the dictatorships and *los desaparecidos* in Latin America. In this regard, in an illuminating study about the function of violence, evil and memory in the work of Roberto Bolaño, Carlos Burgos offers an analysis of the role of the disappeared which can be interestingly applied to *2666*.[65] Burgos defines 'the disappeared' as a category that indicates a simultaneous absence and presence, hence it is always incomplete and open. In addition, the disappeared is a category which brings with it a demand for an explanation: about the nature of the deaths, about the identities and locations of bodies. However, what is most revealing about Burgos' analysis, with respect to *2666*, is that he highlights that even a single death and disappearance is part of a collective issue, since the deaths occurred in the context of mass kidnapping and murder during the dictatorships. In a similar manner, the two sets of disappearances in *2666* are part of Santa Teresa's wave of homicides and, by the end of the novel, part of the Holocaust. Therefore, despite important historical particularities and differences, in the case of the disappeared, as in that of the victims of Santa Teresa's spate of murders, and the Jews exterminated in the Holocaust, the death suffered by a single body, 'adquiere una dimensión colectiva' [acquires a collective dimension].[66] I would like to add to Burgos' analysis by pointing out that the collective character of the crimes in *2666* applies to both perpetrators and victims, since the crimes seem to be the result of a collective organization and affect a whole social spectrum. That said, I would like to sound a note of caution against making too strong a

64 Bolaño, *2666*, 444, 353–354.

65 Carlos Burgos, 'Roberto Bolaño: la violencia, el mal, la memoria', *Nuevo Texto Crítico*, Vol. XXII, 43–44 (2009), 123–144.

66 Burgos, 'Roberto Bolaño: la violencia', 136.

connection between Santa Teresa murders and *los desaparecidos* for, as Jean Franco warns, the femicides 'unlike the disappeared of the Southern Cone [...] were not deaths decreed by the state but "privatized" deaths in an area that has a shadow state'.[67] However, each of the Santa Teresa crimes is part of a collective web, a fact that renders them a social and systemic issue nevertheless. The problem of identification of the bodies, and the will to find a 'solution' to the crimes is, consequently, a social concern.

In his article 'El tono del horror: *2666* de Roberto Bolaño', Carlos Walker observes a close relationship between the reports used in Bolaño's novel to document the femicides and Foucault's description of the clinical gaze. In Walker's view, the forensic reports provide us with 'una articulación entre aquello que se ve y la forma en que se lo describe [...] supone que todo lo visible es enunciable, y que ello es visible por ser íntegramente enunciable' [a joint between what is seen and the form in which it is described [...] that supposes that everything visible is enunciable, and the reason it is visible is because it is enunciable].[68] At the same time, Walker highlights the descriptive rigor and the neutrality of Bolaño's forensic tone which, for Walker, in fact hides the violence of the crimes. This is particularly significant: that clinical language, which generally aims for clarity, should be seen as hiding meaning. Moreover, extending Walker's analysis, the use of the forensic register to account for the murders committed in Santa Teresa carries with it a sense of inadequacy in relation to the literary. The literariness of the novel here seems to be breaking its own protocols, justifying the adoption of an extra-literary and legal tone, thereby questioning its own aptitude to offer a report of the crimes. Alberto Moreiras claims in this regard that 'the relation between literature and murder shows thus, in a privileged and remarkable way, the heteronomy, the impropriety of the literary'.[69] For Moreiras, this unsuitability of the literary proves 'the

67 Jean Franco, 'Questions for Bolaño', *Journal of Latin American Cultural Studies*, 18 (2009), 207–217 (214).

68 Carlos Walker, 'El tono del horror: *2666* de Roberto Bolaño', *Taller de letras*, 46 (2010), 99–112 (106).

69 Alberto Moreiras, 'Infrapolitics and the Thriller: A Prolegomenon to Every Possible Form of Antimoralist Literary Criticism. On Héctor Aguilar Camín's *La guerra*

ethical impropriety of the literary, as literature must yield its autonomy to the pressure of ethics'.[70] And for Moreiras that very connection with the extra-literary, which in *2666* is also linked to the 'secrets of the crimes', is what determines the form of the novel. In Moreiras' words:

> To recognize, to unconceal, to show the secret of the world, if that is the essence of the literary, is an extraliterary essence. The relation between literature and murder seems to posit that the essence of the literary is not itself literary, that the literary apparatus deploys its potentiality at the service of something other than itself. That something other – the secret of the world – determines the structure of the literary apparatus.[71]

Thus, the extra-literariness of the novel in its relation to crime gives the reader the impression that its very constitution takes the form of a secret. For Moreiras, the novel yields 'its autonomy to ethics' through the use of crime fiction which necessarily constitutes a 'political apparatus that seeks to give response to an ethical suspension',[72] the latter being the crimes themselves.

There are further ways in which *2666* connects the notion of the secret, crimes and the literary through genre. In its treatment of multiple crimes: murder, rape, drug trafficking and church desecration, Bolaño's novel makes use of the genre of crime fiction. However, it skews what Franco Moretti has seen as 'a good rule in detective fiction' (a rule which mostly applies to the classic variant of this genre) which consists of having 'only one criminal'. For Moretti, 'the perfect crime – the nightmare of detective fiction – is the featureless, deindividualized crime that anyone could have committed'.[73] In the classic type, crime (what Moreiras defines as an ethical suspension) is commonly presented as an exception, as it is a differentiated individual who perpetrates it.[74] In *2666* the identity of any one criminal is never con-

---

De Galio and *Morir en el Golfo*', in *The Ethics of Latin American Literary Criticism: Reading Otherwise* (New York: Palgrave Macmillan, 2007), 147–179.

70 Moreiras, 'Infrapolitics', 169.

71 Moreiras, 'Infrapolitics', 165.

72 Moreiras, 'Infrapolitics', 170.

73 Franco Moretti, 'Clues', in *Signs Taken for Wonders: On the Sociology of Literary Forms* (London: Verso, 2005), 135.

74 Moretti, 'Clues', 135.

firmed, the only information the reader is given is that there may be several serial killers and that the murders of women are part of an organized and collective wave of crimes. Hence, the collective character of the assassinations challenges society in a way that crimes committed by one person or a couple of individuals would not, in that it strongly suggests that the guilt here is not individual but shared or social. As Moretti remarks, the figure of the criminal is that which must be eradicated from the group, for 'the habitual criminal is an inadequate individual', frequently represented as sick, dangerous or a contaminant, the criminal must be differentiated and excluded from society.[75] Following Moreiras, we could add to this last point by noting that the criminal is also a person who suspends his or her ethical relation to society. However, if the root of the crimes in *2666* is social, the possibility of extracting the criminals is nearly impossible and the solution to criminality has to be of a different kind to the standard identification and exclusion of the criminal.

As a result of these considerations, the response that *2666* offers to the crimes (as constant and social ethical suspensions) is thus partly apocalyptic. But at times the novel also adopts a revolutionary and humorous tone with regards to the crimes. I explained at the beginning of this chapter that the novel established connections between the crimes and advanced legal and illegal capitalism. The idea of a radical movement or a revolution, that would overturn that situation and act as a remedy, crops up with a facetious tone in the novel, when the character María Expósito meets two students in the desert who 'hablaban, por ejemplo de una nueva revolución, una revolución invisible que ya se estaba gestando pero que tardaría en salir a las calles al menos cincuenta años más. O quinientos. O cinco mil' [talked, for example, about a new revolution, an invisible revolution that was already brewing but wouldn't hit the streets for at least fifty years. Or five hundred. Or five thousand].[76] Despite the glaring lack of precision in the students' predictions, the obvious question is: does this revolution include any kind of optimistic end result?

75 Moretti, 'Clues', 138.

76 Bolaño, *2666*, 697, 558.

The theme of a revolution is taken on again in the last part of the novel when Archimboldi discovers Boris Ansky's notes and finds out his thoughts on the matter: 'la revolución, pensaba Ansky, terminará aboliendo la muerte' [ultimately, thought Ansky, the revolution would abolish death].[77] To which his friend, the writer Ivanov, used to reply that that was impossible, since death was a companion of man since time immemorial.[78] Ansky, in turn, is reported to have replied: 'de eso precisamente se trataba, justo de eso, incluso exclusivamente de eso, abolir la muerte, abolirla para siempre, sumergirnos todos en lo desconocido hasta encontrar otra cosa. La abolición, la abolición, la abolición' [that was precisely it, the whole point, maybe the *only* thing that mattered, abolishing death, abolishing it forever, immersing ourselves in the unknown until we found something else. Abolishment, abolishment, abolishment].[79] Ansky's revolution recalls apocalyptic proceedings particularly in the way he exhorts us to first 'sumergirnos' [immerse ourselves] in the unrevealed in order to root out death and to find something different.

Going back to my initial points concerning the resolution of the crimes in traditional crime fiction, despite the above expressions of revolutionary hope for finding alternatives, the restitution of the ethical relationship that the classic detective novel manages, in solving the crime, is absent in *2666*. Yet it is as if Bolaño's novel rehearsed different responses, apocalyptic, humorous and revolutionary, to that 'tradition that restitutes a sinister heteronomy at the heart of the literary'.[80]

In a subtler manner, the crimes and the dead bodies of women are also alluded to in the novel's comparison of the desert to a garden. This association appears in the farewell to the scholars by the chairman of the Sonoran

77 Bolaño, *2666*, 890, 712.

78 A relevant reference to the Apocalypse that seems to come out of the blue in the novel appears when Ivanov was celebrating the success of one of his books he has dinner at the 'Writer's House' with 'cuatro amigos que parecían los cuatro jinetes del Apocalipsis' [four friends who resembled the four horsemen of the Apocalypse], 890, 712.

79 Bolaño, *2666*, 888, 710.

80 Moreiras, 'Infrapolitics', 172.

University, Augusto Guerra, which 'estaba adornada con una frase poética que comparaba el desierto con un jardín petrificado' [in a poetic turn of phrase, the flowery closing compared the desert to a petrified garden].[81] This analogy unites both spaces and annuls their opposed character. The antagonism between the two landscapes lies in the conception of the desert as a non-productive region devoid of vegetation, and therefore devalued, and the garden as place of abundant vegetation, often used as a metaphor of Eden (as it is 'petrified' here, it is an 'immobilized' or 'dazed' Eden). The image of the desert as a petrified garden not only contains and holds in tension normally opposed spaces, I would argue that this is also an allusion to the stiff bodies of the murdered women and the absence of their gravestones in the desert. For gravestone, in Spanish, can be termed *piedra sepulcral*, and the word 'piedra' shares its Latin etymological root, 'petra', with the word 'petrificado', used in the quotation above to describe the garden. Since there is no 'piedra sepulcral' in that space, it is the garden with the dead bodies which by extension seems to have become 'petrified'.

The garden is significant not only in its comparison to the desert but also in its use as a metaphorical site to explain the novel's poetics. We observe this in the last part of the novel when Archimboldi and another recently appeared character, Pückler, have a conversation in which the latter speaks about his ancestor, a travel writer. According to him, through the description of forgotten gardens 'dejados de la mano de Dios' [abandoned to their fate],[82] Pückler's ancestor disclosed an accurate image of his time – the Europe of the moment – and of the conditions of humankind. From a garden that is in the middle of the 'maleza' [weeds] and 'desidia' [slovenliness], that writer, Bolaño tells us, was able to *unveil* 'las vicisitudes de la condición humana' [the vicissitudes of the human condition].[83] Pückler highlights the perplexing nature of his ancestor's work and qualifies it as 'vaya legado más misterioso ¿no cree usted?' [a mysterious legacy, don't you think?].[84] The encounter with Pückler is key not only because it closes the

81 Bolaño, *2666*, 151, 112.
82 Bolaño, *2666*, 1118, 892.
83 Bolaño, *2666*, 1119, 893.
84 Bolaño, *2666*, 1119, 893.

novel, but because it registers what I understand is Bolaño's declaration of aesthetic principles: *2666* contains veiled meanings and the reading of it becomes disclosure. This last view is underpinned elsewhere in the last part of the novel when another writer, who sells the typewriter to Archimboldi, states that a minor writer (who is also capable of writing masterpieces) achieves a decent work 'no por un ejercicio de estilo o voluntad [...] sino gracias a un ejercicio de *ocultamiento*' [not from an exercise of style or will [...] but as the result of an exercise of *concealment*].[85] Bolaño's literary proposal also includes concealment (like Pückler's) for it seems to have the expectation of proposing a revelatory reading. The closure of the scene with Pückler's relative (which is also the closure of the book) is also a poignant allusion to the Apocalypse. After absolute darkness, the lights come on in the park: 'De pronto se encendieron las luces del parque, aunque hubo un segundo de oscuridad total, como si alguien hubiera arrojado una manta negra sobre algunos barrios de Hamburgo' [Suddenly the park lights came on, although there was a second of total darkness, as if someone had tossed a black blanket over parts of Hamburg].[86]

Writing time: 'treasure hidden in a desert cave'

At the beginning of this chapter I indicated how notions of space and time are fundamental to my reading of *2666*. One of the central figures to bring matters of space and time to currency in the novel is Amalfitano. When Amalfitano randomly finds a book in his house, the *Testamento geométrico*

85 Bolaño, *2666*, 983, 786 (emphasis in the original). This character later adds a mysterious comment before Archimboldi leaves: 'Jesús es la obra maestra. Los ladrones son las obras menores. ¿Por qué están allí? No para realzar la crucifixión, como algunas almas cándidas creen, sino para ocultarla' [Jesus is the masterpiece. The thieves are minor works. Why are they there? Not to extol [*sic*] the crucifixion, as some innocent souls believe, but to hide it], 989, 790.

86 Bolaño, *2666*, 1119, 893.

by Rafael Dieste, which he does not recognize as being his property, his perplexity is used to offer a detailed review of what the book contains. Brett Levinson notes that the very name of Dieste's book is a possible indication that we are about to discover Bolaño's 'testimony' (*testamento*) on form (*geométrico*).[87] Through Amalfitano's reading, Bolaño describes the organization of the book and quotes the different sections which deal with Euclidean and non-Euclidean geometry. He also mentions that the book not only contains the author's thoughts and investigations about space but also that Dieste was a mathematician and a poet. Here, I believe, the reader is being asked to pay attention to the different ways in which space can be conceptualized, its dependence on geometry and its relationship with literature.

In the description of Dieste's book it is noted that the prologue is entitled 'la intuición iluminada' [Enlightened Intuition] and that the person who wrote it 'hasta se concedía el lujo de citar a Heidegger (*Es gibt Zeit: hay tiempo*)' [went so far as to quote Heidegeer (*Es gibt Zeit: there is time*)].[88] This phrase is particularly relevant since the statement 'hay tiempo' is of course a colloquial way of saying that there is time to use, to live, to do, before something else. Likewise, that the author of the prologue is described as having allowed himself the luxury (se concedía el lujo) of using the quote from Martin Heidegger to say that there was time, for he was using (or wasting) part of the time there was to affirm that there was time left, constitutes an ironic note on the issue. What is fundamental here is that *Es gibt Zeit: there is time* alludes to a philosophical concept of time developed by Heidegger in *Being and Time* (1927), which, as Lilian Alweiss explains, 'questions the whole tradition of Western thought which, ever since Plato, has viewed time as "a movable image of Eternity"'.[89]

Heidegger sets against the common concept of time (the time of the calendar and social time) what he calls 'temporality' which is based on the

87 Brett Levinson, 'Case Closed: Madness and Dissociation in *2666*', *Journal of Latin American Cultural Studies*, 18 (2009), 177–191.

88 Bolaño, *2666*, 244, 189.

89 Lilian Alweiss, 'Heidegger and "The Concept of Time"', *History of the Human Sciences*, 15 (2002), 117–132 (119).

duration of human life-span and is therefore subjective. Heidegger questions traditional ways of seeing time in relation to eternity (as for instance in Plato and Christianity) since for Heidegger 'we can only understand the phenomenon of time from our mortal or finite vantage point' and not from eternity.[90] We can never estrange ourselves from our temporal existence, since 'eternity as a point of departure is never at our disposal'.[91] Therefore, contrary to traditional philosophy and its understanding of time from eternity, Heidegger supports a view of time as a concept that finds its meaning from death. This concept of time is, I believe, also fundamental to the understanding of *2666* since in this novel Bolaño seems to be thematizing his own closeness to death. The 'hay tiempo' of the novel refers then to this Heideggerian concept of time which has incorporated death into itself. Death for Bolaño, as for Heidegger, becomes the limit from which to trace the path to the beginning, or the start. A reference to this concept of time crops up again in the last part of *2666* when the writer who sells the typewriter to Archimboldi describes a conversation he had with the employee of a morgue:

> Y su respuesta: *no tengo mucho tiempo*. ¡Cuántas puertas abría esa respuesta! ¡Cuántos caminos quedaban de pronto despejados, visibles, tras esa respuesta! [...] No tengo mucho tiempo, estoy viviendo. No tengo mucho tiempo, me estoy muriendo. Como usted comprenderá, ya no hubo más preguntas.
>
> [And his reply: *I don't have much time*. How many doors it opened! How many paths were suddenly cleared, revealed to me! [...] I don't have much time, I'm busy living. I don't have much time, I'm busy dying. As you can imagine, there were no more questions.][92]

The employee of the morgue says he does not have much time, implying therefore that he has some time. The time that he has (which is always not much) is measured in terms of life and in terms of death. Thus, death is to be faced up to while we live. This brings to mind David R. Cerbone's

90 Alweiss, 'Heidegger and "The Concept of Time"', 117.

91 Quoted in Lilian Alweiss' 'Heidegger and "The Concept of Time"', 120.

92 Bolaño, *2666*, 988–989, 790 (emphasis in the original version and the translation).

insightful observations concerning Heidegger's *Being and Time*, 'Death thus makes vivid the idea that I *have* time, that there is a time that is "mine" to be *used* [...] only a being who can acknowledge its mortality in this way can bear this sort of relation to time'.[93]

Furthermore, for Bolaño, the acknowledgement of this relationship between time and death also emerges from suffering an illness. Thus, when the novel presents Archimboldi's view on the type of notions of time that ill and healthy people have, it is remarked that: 'La noción del tiempo, ah, la noción del tiempo de los enfermos, qué tesoro escondido en una cueva en el desierto' [That [*sic*] sense of time, ah, the diseased man's sense of time, what treasure hidden in a desert cave].[94] We can explain that quotation, I believe, returning to Heidegger's philosophy. In *The Essence of Truth* Heidegger analyses Plato's allegory of the cave in order to prove that truth is not a correspondence or a proof of correctness but disclosure. Heidegger uses the Greek word *Aletheia*, which means 'unconcealment' or 'deconcealment'. Heidegger explains that:

> Only if beings are previously experienced in their hiddenness and self-concealment, if the hiddenness of beings encompasses man and besets him in a fundamental way, only then is it necessary and possible for man to set about wrestling beings from hiddenness and bringing them into unhiddenness.[95]

Following Heidegger, the notion of time here is concealed in the cave, in order to be finally un-hidden. Hence, to read time and about time means to un-hide and disclose its meaning from the cave. It should be noted, however, that depictions of time in *2666* are not always the same nor do they direct us to the same meanings: they meander from representations of time as final, as insufficient, as something man wishes to rule over, to time as a path to decay and finally to time as a way to liberation. Some of these views are also paralleled in the ways the Apocalypse has been read, as

93 David R. Cerbone, *Heidegger: A Guide for the Perplexed* (London and New York: Continuum International Publishing Group, 2008), 86.

94 Bolaño, *2666*, 825, 661.

95 Martin Heidegger, *The Essence of Truth: On Plato's Cave Allegory and Theaetetus* (London: Continuum, 2009), 9.

mentioned at the beginning of this chapter, since it has been interpreted as a text of consolation for the struggles of the present and promising redemption at the end, or as a grand destructive finale.

I would like to turn to two instances in the novel in which we observe examples of time conceived either as as liberation or as decay. The first of these instances involves a dialogue between two characters, Fate and Chucho Flores. Flores summarizes the pros and cons of the city of Santa Teresa, including the maquiladoras, very low unemployment rates, money, poverty, a cocaine cartel and so on, but in his opinion, the only thing they are missing is time: '– Falta el jodido tiempo. ¿Tiempo para qué?, pensó Fate ¿Tiempo para que esta mierda, a mitad de camino entre un cementerio olvidado y un basurero se convierta en una especie de Detroit?' ['We haven't got any fucking time'. Time for what? thought Fate. Time for this shithole, equal parts lost cemetery and garbage dump, to turn into a kind of Detroit?].[96] More time, felt as a need for Chucho Flores, is seen as a clear path to a greater decline for Fate.

The second instance also concerns a conversation, this time between the main suspect of the femicides, Hass, and the character Sergio González. Haas is in jail and awaiting trial, and during this conversation he confides that he knows there must be people who definitely know he is innocent, but to find out exactly who those individuals were: 'necesito tiempo, lo que en mi caso resulta paradójico, ¿no le parece? ¿Por qué? Dijo Sergio. Pues porque aquí lo único que tengo en abundancia es tiempo. Pero yo necesito más tiempo aún, mucho más, dijo Haas' [I need time, which in my case is paradoxical, don't you think? Why? asked Sergio. Because all I have here is plenty of time. But I need even more time, much more, said Haas].[97] For Haas, more time is translated as the way to liberation from jail. He is however stuck in a puzzling paradox, part of which he identifies in the quote – that he simultaneously needs more time but also that he has nothing but time – the other part being that more time could lead to either of two different 'apocalyptic' outcomes, a condemnation or liberation.

96 Bolaño, *2666*, 362, 286.
97 Bolaño, *2666*, 614, 490.

I would now like to round up the ideas described in this chapter. Contrary to the reviewed studies which talk of the impossibility of meaning or the cancelation of sense in Bolaño's novel, all the representations related to the desert closely read here speak to the significance that the desert hides meanings that could be unhidden. Examples of this seen in this chapter are the author's descriptions of dreams in which the desert appears and which contain meaning, both concealed and manifest, and the women who 'disappear' in the desert whose identities could be determined only if they were to be unburied and examined. The desert, likewise, is not a space characterized in a traditional way by its emptiness, since it is through the reading of this space that the notion of time, dreams, and memories are made suitable to disclosure. Moreover, the desert not only exposes the relationship that the novel has with the Apocalypse; it also performs a cohesive function since it unites the different parts of *2666*.

Other aspects of the novel further support the thesis of *2666* as a repository of (hidden) meaning: the title, that 'microcosm' of the narrative encoded in a number that has to be disclosed in order to apprehend some of its significance; and the crimes which relate the novel to the genre of crime fiction, which is itself, as Moretti points out, 'a double system of meanings, superficial and deep: the first is both the manifestation and the cover for the second'.[98] Moreover, though with varied meanings, space and time are held in the novel to be read in terms similar to those of the Apocalypse and dreams: they are to be disclosed and their meanings are both hidden and un-hidden. This can lead to the conclusion that polyvalence and ambiguity play an important role in Bolaño's novel. While my chapter does not deny that, I would like to highlight that *2666* also proposes reinstating meanings as a form of reading, independent of how polyvalent those meanings are. The analysis of time put forward in this chapter is the clearest example of a vital notion that conveys meanings, while swaying to and from different perceptions and conceptions of time.

I hope that my elucidations within this chapter have evidenced the intention to give the desert and the notion of the secret a more adjusted

98 Moretti, 'Clues', 134

existence to the novel's aesthetic proposal than the idea of complete indeterminacy and lack of meaning allows for. In addition, although *2666* clearly lacks any denunciatory tone, I hope that I have also been able to show the different ways in which the novel is concerned with ethical problems posed by the femicides and the issues of an all pervasive narco-culture and extreme capitalism in a peripheral terrain. Earlier I detailed Poblete's formulation of how death marks the end both of this novel and of Bolaño's life. I would like to add to her formulations that, as in Heidegger's notion of temporality, death is something that the novel faces up to, not only as the finale waiting at the end of the novel, but as a constant rhythm of the narrative. Concerning the novel's ending, here we find an important unanswered question: '¿por qué una obra maestra necesita estar oculta? ¿Qué extrañas fuerzas la arrastran hacia el secreto y el misterio?' [why does a masterpiece need to be hidden? what strange forces wreath it in secrecy and mystery?].[99] I would contend that, perhaps, Bolaño sets his novel in the terrain of mystery so that the interest and fascination found in revealing that mystery might act as a possible answer to death, or as a recipe against forgetfulness. Alternatively, as Bolaño expresses through the character Pückler (the German writer whom Archimboldi meets at the end of the novel), with *2666*'s 'exercise of concealment' Bolaño could also be trying to propose revelatory and meaningful readings.

99 Bolaño, *2666*, 984, 786.

CHAPTER 5

The Event of Literature: A Reading of the Desert and Migration in Juan José Saer's *La ocasión*

De esa imposibilidad de avanzar, de llegar a alguna parte, surge una característica muy importante en el Quijote, que son los fuertes acentos de melancolía y de fracaso que impregnan toda la obra. Don Quijote, Caballero de la Triste Figura, es como ya lo sabemos un personaje doliente y una de las causas más importantes del impacto que produce en nuestra imaginación es que resulta difícil estimar hasta qué punto Don Quijote es consciente de su situación singular.[1]

In the epigraph taken from one of Saer's collections of essays, the author singles out three main experiences that take place in *El Quijote*: the intention to change places, to get somewhere, both intellectually and physically; the resulting feelings of those attempts, namely failure and melancholy; and the difficulties of apprehending the modulations of the main character's consciousness (given his difficulties in discerning between reality and illusion). For Saer these experiences are key and appear time and again in modern literature since Cervantes. Some of the emblematic examples that Saer offers in his essay are Laurence Sterne's *Tristram Shandy* (1759), Gustave Flaubert's *Madame Bovary* (1856) and James Joyce's *Ulysses* (1922). Each of these authors writes with the influence of Cervantes's text and though their novels are different in content, scale and time of writing, in all of them we find those fundamental experiences from Cervantes's text identified by Saer. In the same manner, I believe that Saer himself draws on the ideas he formulated about *El*

1 Juan José Saer, 'Líneas del Quijote', in *La narración objeto* (Buenos Aires: Seix Barral, 1999), 31–54 (45).

Quijote to write his 1988 novel entitled *La ocasión*: a text centred on the story of an immigrant who moves to the Argentine Pampas ('the desert'); an account in which failure, the intricacies of consciousness, and both the importance of moving from place to place and its effects on identity are essential elements of the novel.[2]

Part of the significance of the epigraph above for the reading of *La ocasión* also lies in Saer's underlining of the uncertainties for the reader provoked by Cervantes's text; for he or she struggles to gauge the exact extent to which Don Quijote 'es consciente de su situación singular' [is conscious of the peculiar situation he is in]. What is foregrounded in the epigraph is that the unsureness of the text also becomes that of the reader through 'el impacto que [el texto] produce en nuestra imaginación' [the impact [that the text] produces in our imagination]. Indeed, considering that impact one can also say that, to a great extent, the text's value is embedded in that 'experience' of reading: in that 'ocasión', or event. Conceived as such, the performative meaning of the literary text is especially relevant in the epigraph above and, I believe, to the reading of *La ocasión*, whose very title seems to suggest from the outset that works of literature 'are processes of signification materialized only in the practice of reading'.[3] In line with this notion, Derek Attridge explains in *The Singularity of Literature* that the work of literature is always an event since the reader encounters and undergoes the text, being exposed to it and transformed by it.[4] While Attridge highlights the importance of the encounter between text and reader, he also gives form a pivotal role in the consideration of literature as an event, remarking that 'the literary use of language involves the performing of meanings and feelings, and that what has traditionally been called

2 Juan José Saer, *La ocasión* (Barcelona: Círculo de lectores, 1988). All further references are taken from this edition, subsequent page references in the text refer first to the original and then to the translation in English: *The Event*, trans. Helen Lane (London: Serpent's Tail, 1995). I should note that this chapter is not an intertextual reading of Saer's novel with *Don Quijote de la Mancha*, nor a reading of its influence per se.

3 Terry Eagleton, *Literary Theory: An Introduction* (Oxford: Blackwell Publishers, 1983), 74.

4 Derek Attridge, *The Singularity of Literature* (London: Routledge, 2004).

form is central to this performance'.[5] The centrality of the role played by form is a constituent of what Terry Eagleton in *The Event of Literature* has called 'one of the paradoxes of the literary work', since 'it is "structure" in the sense of being unalterable and self-complete, yet "event" in the sense that this self-completion is perpetually in motion, realised as it is only in the act of reading'.[6] The work of literature thus cannot be reduced to its contextual or formal reality, it interacts with the reader in the process of meaning. This is not to say that the text exhausts itself in the performative act of reading, for the work of literature is, as Eagleton points out, 'by virtue of its performative statements, caught up in a network of significations which exceeds and sometimes subverts that performance'.[7] Moreover, the consideration of the literary text as engaged in its own enactment does not presuppose an aesthetically isolated and self-contained interpretation, but one that is affected by history and context. For that reason, Edward Said points out that any literary text is, to some extent, an event because works of literature 'are worldly [...] even when they appear to deny it, they are nevertheless a part of the social world, human life, and of course the historical moments in which they are located and interpreted'.[8] Thus, in their different considerations of literature as an event, Saer's comments, along with the reflections of Said, Attridge and Eagleton make a common point in stressing the importance of the *experience* of reading as a worldly event. Within this frame of thought, we can state that reading, as Said pointed out, brings us 'into an awareness of how language is about experience and not just about itself'.[9] Thus, I believe that both the epigraph

5 Derek Attridge, *J. M. Coetzee and the Ethics of Reading: Literature in The Event* (Chicago and London: The University of Chicago Press, 2004), 9.

6 Terry Eagleton, *The Event of Literature* (New Haven, CT and London: Yale University Press, 2012), 201.

7 Terry Eagleton, 'A conversation with Terry Eagleton', interview with José Manuel Barbeito Varela, *Atlantis*, vol. 23, nº 2 (2001), 169–185, 176.

8 Edward Said, *The World, The Text, and The Critic* (Cambridge, MA: Harvard University Press, 1983), 4.

9 Edward Said, *Reflections on Exile and other Literary and Cultural Essays* (London: Granta Books, 2001), xv. Said specifically notes that this is the reading he makes of works of exiled writers such as Conrad, Nabokov, Joyce and Ishiguro.

to this chapter and the title of Saer's novel invite us to consider reading as experience, as an event. Therefore, in the sections that follow, I start by examining the experiences foregrounded by Saer's fiction, of which travel is the central one. If commonly a potent metaphor for achievement and gain in modern texts, travel in Saer's work introduces the case of necessity, estrangement and failure. Drawing on Caren Kaplan's work, I examine the novel's shaping and understanding of travel as demystified displacement. From there I move on to how *La ocasión* signals the importance of migration by both giving an account of the transformations of a character through travel, while at the same time formally distancing the reader from the narration. Accordingly, I will demonstrate how *writing travel* in *La ocasión* echoes in form the experience of distancing produced by migration while foregrounding the mindfulness of reading as an event in itself.

Writing travel and the poetics of failure

The plot of *La ocasión* hinges on tensions formed around the journeys of the protagonist and the transformations of his identity. In this way, the novel foregrounds numerous individual events whilst also reflecting upon the repercussions that the continual change of place has on identity and how those effects are translated textually. The novel's main character, Bianco, is a telepath, a dilettante in the occult who believes he has special powers which would prove wrong the positivism rampant in Europe in the second half of the nineteenth century. He therefore declares on a few occasions in the novel to have repeatedly demonstrated that he could dominate matter with his mind. As with *Don Quijote*, we do not know to what extent he is 'consciente de su situación singular' [conscious of the peculiar situation he is in], but we know that it is failure (at being able to demonstrate his powers) that impels his ensuing migrations. These circumstances reach their climax at a high-profile *event* organized with the intention of consecrating Bianco's fame. There, he is ambushed by a fellow ex-mysticist-magician (who has also been a victim of entrapment like Bianco) and a group of positivists.

During this show they ridicule him by having a clown perform the same demonstrations of special powers as Bianco (bending metals, making old broken clocks work again without touching them and so on), with the result that the whole audience laughs at him and turns against him. After this, his name is sullied and his new fame as a charlatan spreads throughout several European countries. He flees first to Normandy and then to Sicily, where he remains full of anger, paranoid that people recognize him everywhere as the subject of the great scandal. Almost randomly, Bianco finds a job recruiting immigrants from Italy to settle in Argentina and he ends up travelling with them to live in the open Pampas. As he remains profoundly embittered about the incident in which he was ambushed by the positivists, he aims to write a piece which will refute them and their ideas. In Argentina he immediately befriends a young doctor, Antonio Garay López, who also becomes his business partner and neighbour, as Bianco acquires acres of land adjacent to the doctor's family territories. Through his new position as a successful land owner and entrepreneur he manages to arrange a marriage with Gina, the beautiful, young daughter of some local workers (although originally Italian). The novel finishes after Bianco and Gina (who is expecting a child) escape the yellow-fever epidemic in the province of Buenos Aires, moving to Bianco's basic and secluded homestead in the middle of the country-side.

At first sight, the outline of the plot I have offered above could be seen as ascribing the protagonist to the modernist figure of the exile, who, though estranged from his past and familiar location, makes progress out of the distance forced upon him or her and is able to construct or obtain some kind of 'gain through exile'.[10] Not only does Bianco become financially successful in Argentina, in fact, achievement through travel even acquires a messianic veneer for Bianco. Humiliated in Paris, because a group of people did not believe in his special endowment, having suffered humiliation at the hands of unbelievers, he decides to retire for a period of time to the 'llanura-desierto' [plains-desert] in order to be able to 'come back' with the written word which will liberate mankind from matter: 'él

10 Caren Kaplan, *Questions of Travel: Postmodern Discourses of Displacement* (Durham, NC and London: Duke University Press, 1996), 36.

debía perderse otra vez, durante años, en la penumbra, y resurgir con esas páginas luminosas' [he needed to lose himself in the shadow again for several years and then reemerge [*sic*] with these luminous pages].[11] Thus, on the face of it, the story of migration in *La ocasión* seems to correspond with what Caren Kaplan describes in her study of discourses of travel in modern and postmodern writing: 'exile becomes the paramount model for the production of modern literature'.[12] That paradigm relies heavily on the figure of the artist as a displaced person and on the 'sense of "exile" as authorship'.[13] Nonetheless, as Kaplan warns us, that model easily lends itself to being complacent and celebratory about exile participating 'in the mythologized narrativizations of displacement without questioning the cultural, political, and economic grounds' of travel.[14] Kaplan claims that seeing accounts of displacement with a lens more carefully adjusted to the 'personal and national histories' of the people who migrated, allows us to become aware that their lives and works 'are far more complex than a "celebration of vagabondage" would allow us to consider'.[15]

On further consideration, however, the plot of *La ocasión* and its *events*, do not seem to support the model of exile resulting in gain (creative or not) and therefore do not ultimately reinforce the view of travel as 'celebration'. This can be seen in the fact that almost every life story in *La ocasión* is marked not only by the events themselves (journeys, sojourns and also missed opportunities) but also by the disillusionment, loss and failure that looms from them. The most concrete example of this is of course the foremost defining moment in the novel when Bianco appears at what he thinks is going to be his greatest moment, a demonstration of his control over matter, only to suffer an orchestrated humiliation by the Positivists, made all the more hurtful because they use a fellow mystic as their instrument. Bianco's planned work to rebut positivism never comes to fruition; neither does an occasion ever arise for him to demonstrate the

11 Saer, *La ocasión*, 35, 18.
12 Kaplan, *Questions of Travel*, 39.
13 Kaplan, *Questions of Travel*, 39.
14 Kaplan, *Questions of Travel*, 2.
15 Kaplan, *Questions of Travel*, 39.

falsehood of the positivists who humiliated him. Another instance of a failed undertaking occurs when Bianco finally finds the opportunity to ask his business partner and friend in Argentina, Garay López, a particularly difficult question, that is, whether Garay López and Bianco's wife have had an affair. This question tortures Bianco from the point in the novel where he encounters Garay López and his wife together in the house and, unfortunately, Garay López dies of yellow fever before Bianco can ask him about the suspected affair, leaving his question unanswered. Thus, the novel finishes without clarifying Bianco's distressing doubt, a state of uncertainty that the character also articulates by asking himself a further question concerning whether the baby that his wife, Gina, is carrying will have Garay López's black hair or be a read-head like himself – indicating that he is the father (although there is the obvious possibility that the baby would have his/her mother's black hair, the main character doesn't even consider this). This state of despair, caused by the lack of an answer to a fundamental question or quest, is echoed in a fragment in the second chapter, in which the character Garay López reads out loud his 'alegoría teatral' for Bianco's comment. The allegory is based on the biblical story of the three kings who follow the star to Bethlehem to find Jesus. In Garay López re-telling, the kings arrive in town after a long and less than harmonious trip following the star, in the company of a band of fellow seekers who joined them along the way. Once in the town they ask for the baby who was just born. The locals reply that no babies have been born that day, or even recently, and although the new arrivals search everywhere, helped generously by the inhabitants, there is in fact no baby to be found. The story appears to be an allegory for Bianco's own journey, and like him, the kings and their followers become disappointed due to the futility of their journeys and the lack of knowledge or certainty as to what to do next. Later in the novel there is another echo of this doubt and indecisiveness, resulting from an unfulfilled opportunity. In this case we learn of a Calabrian immigrant who is waiting anxiously for the opportunity to speak with a clairvoyant boy called Waldo in order to ask if he should go back to Italy or remain in Argentina. The Calabrian eventually manages to see the boy but, the room is full of people and he leaves very disappointed as he is not able to understand the boy's words (his 'predictions' or 'prophesies' are cryptic and given in mumbled, repeated rhyme and the Calabrian is not fluent in Spanish). As a result he is left in a state of profound sadness without any clearer idea of what decision to make. Here, as described in the other cases, it is not gain but failure that underlies events. Thus, in *La ocasión*, Saer overturns the idea of exile (and of other arduous journeys and quests) as gain, instead adhering to a poetics of failure that, as I indicated at the start of this chapter, reaches back to Cervantes.[16]

Changing selves, travel and distancing

Florencia Abbate argues that *La ocasión* narrates the transformations that took place in the province of Santa Fé, Argentina, during the the waves of immigration which occurred in the second half of the nineteenth century. This scholar points out that, in that precise feature, Saer's text seems to emulate historical novels of the nineteenth century, whose aim was to define the 'ser nacional' [national being]. Abbate considers that, despite similarities between *La ocasión* and nineteenth-century texts (which Abbatte also

16 For more on this, see Saer's 'Líneas del Quijote', in *La narración* (1999), 31–54 and 'Nuevas deudas con el *Quijote*', in *Trabajos* (Buenos Aires: Seix Barral, 2006), 79–82. In this last essay Saer notes that Cervantes's 'moral of failure' is a feature of modern Western literature: 'A diferencia del héroe épico, que espera un progreso como resultado de sus aventuras [...] Don Quijote se encuentra al final de cada una de las suyas en el mismo lugar, defraudado e incluso malherido, física y moralmente, y sin embargo, aun habiendo anticipado vagamente su fracaso, decide continuar sus aventuras. Tal es la moral del fracaso que inaugura Don Quijote de la Mancha, y que está presente en la casi totalidad de la narrativa occidental moderna' (80) [Unlike the epic hero, who expects progress as a result of his adventures [...] Don Quixote finds himself in the same place at the end of each of his adventures, disappointed and in fact seriously wounded, physically and morally. Nevertheless, even though he had vaguely anticipated his failure, he decides to continue with his adventures. Such is the moral of failure that Don Quixote de la Mancha inaugurates, and which is present in almost the totality of modern Western literature] (my translation).

proposes include the omniscient narrator and the emblematic nature of secondary characters), Saer's text does not determine but rather problematizes identity. As Abbate puts it, *La ocasión* 'no descubre en los orígenes ningún "ser" nacional. Antes bien, revela que en la raíz de la identidad nacional no se encuentra ninguna esencia sino más bien la exterioridad del accidente' [does not discover in the origins a national 'being', but reveals that in the root of national identity there is no essence but the exteriority of accident].[17] I would like to add to Abbate's reflections that, to a great extent, what she calls the 'exteriority of accident' mainly hangs on the main character's lack of a coherent narrative for his past and his continuous journeys or migrations, both features which also distinguish Saer's novel from those nineteenth-century novels invested in the demarcation of a 'national being'. Moreover, in the problematization of the main character's identity as noted by Abbate, *La ocasión* not only relies on 'the exteriority of accident', but also relates to Freudian ideas of the self, since Bianco is never entirely an owner of his own destiny, his agency never fully dependent on his self-will and never completely the result of his conscious desires. Furthermore, if *La ocasión* explores the intricacies of the conscious and unconscious mind of the protagonist, it also relies heavily on the idea that 'travel has long been a means of changing selves'.[18] Accordingly, Saer's novel can be seen not only as centred around 'the exteriority of accident', but also the interplay of consciousness and the unconscious, and the stresses that persistent travelling effects on the mutability of an individual.

La ocasión continuously informs the reader of the shifting nature of the main character's identity and his numerous migrations. For instance, the book opens with an explanation about the indecisiveness of the name of the main character, who is 'normally' called Andrea Bianco, but also Andrew Burton, or just A. Bianco. Some of the reasons for changing names sound like farfetched excuses, such as using the surname Burton because the name Bianco 'puede minar la credibilidad de un pelirrojo' [might ruin the

17 Florencia Abbate, 'La crisis de la identidad en el relato histórico: un análisis de *La ocasión* de Juan José Saer', *Amerika: Memoires, identites, territories* 9 (2013), 78–89 (89).

18 Eric Leed, 263.

credibility of a redhead].[19] The protagonist stops using the name Burton after the dispute with the positivists in Paris and at one point Andrea Bianco-Andrew Burton combines both surnames before following the suggestion of a young Prussian and making the decision to use only the name Bianco, 'para neutralizar la desconfianza legendaria del resto de los europeos hacia los ingleses' [so as to escape the legendary mistrust of Europeans vis-à-vis the English].[20] The continual changing of names and locations in *La ocasión* brings to mind Eric Leed's argument that 'the transformations of social being in travel suggest that there is no self without an other; and that, at bottom, identity is done with mirrors'.[21] Yet the main character's changes of name are also suggestive of an identity that is irresolute and multiple. In fact, Saer tells us that there is, in his 'indecisión en lo relativo a su nombre, una resistencia a dejarse representar por uno solo, como si temiese que [...] muchas partes de su ser se secaran y desaparecieran' [uncertainty about his name, a resistance at letting himself be represented by a sole patronymic, as though he was afraid that, by choosing too definite a name, a number of parts of his being would dry up and disappear].[22] As I will explain further below, this type of multiple-identity seems to result from the protagonist's numerous voyages: his high mobility transmutes into a shifting personality. Stuart Hall defines identity as 'always in process and always constituted within not outside representation', but for Bianco the process of constructing an identity is further complicated by his journeys and what the novel calls 'the haziness' of his past.[23] If, as Hall notes, identity is grounded in the 're-telling of the past', that re-telling is an impossibility for the main character of *La ocasión* given that Bianco's origins and past are not clear either to the reader, the narrator or, it seems, even to the protagonist.

Bianco's indecisiveness with regards to his own name is matched by an uncertainty surrounding his origins. To begin with, we are told that

19 Saer, *La ocasión* 19, 1.

20 Saer, *La ocasión*, 25, 7.

21 Eric J. Leed, *The Mind of the Traveller* (New York: Basic Books, 1991), 264.

22 Saer, *La ocasión*, 28, 10.

23 Stuart Hall, 'Cultural Identity and Diaspora', in *Identity: Community, Culture, Difference*, ed. J. Rutherford (London: Lawrence & Wishart, 1990), 222–237 (222).

Bianco used to say he was born in Malta to explain his accent, but that after a few years he stopped this practice. He is not Italian, since it is said that he adopts this nationality later. He lives in several countries, namely Prussia, England, France, Italy and Argentina, though he is not native to any of these. His mother-tongue is also undetermined and it is impossible to guess this from his undefined accent: 'si pudiese saberse con exactitud cuál es ese idioma materno' [providing one could find out exactly what language is his mother tongue].[24] Indeed, his identity and past are characterized time and again in the novel with the same adjective 'brumosa-o' [hazy, misty]: 'pasado brumoso' [hazy past],[25] 'procedencia brumosa' [obscure origins],[26] and 'zona brumosa' [obscure, hazy area].[27] Thus, the novel creates a character whose referential features are deconstructed through an upholding of multiple possibilities acquired through travel.

One question we should ask ourselves, however, in relation to the previous discussion is: who is defining the main character's identity as hazy and who is wondering what his mother tongue is? For although the novel is narrated in the third person and the main character is seen from outside, the narrator is omniscient and takes Bianco's point of view; that is, every person, landscape or situation is narrated and described in terms of how the protagonist thinks or feels about them. The narration acts and reacts to the main character's state of mind. The omniscient narrator, so closely aligned with the main character's thoughts and feelings, makes it all the more unusual that Bianco's origins are described just as hazy and obscure. The interventions of the narrator are, of course, a reproduction of a common feature of nineteenth-century novels from which *La ocasión* draws. In contrast to these texts, however, the intercessions from Saer's narrator are not there to guide the response of the reader or to clarify aspects of the plot. Rather, the characterization of the main character's origins as hazy and the interventions of the narrator are concurrently a literary device employed by Saer to distance the reader and to emphasize that the

24 Saer, *La ocasión*, 84, 72.
25 Saer, *La ocasión*, 65, 51.
26 Saer, *La ocasión*, 19, 1.
27 Saer, *La ocasión*, 34, 17.

text being read is fiction: this is a story about a travelling man of obscure origins (which will not or cannot be clarified even by the omniscient narrator) who thought he had special powers. As such, this distancing effect formally reproduces the process of *estrangement* experienced by the main character in his 'changes of self' through travel and the unaccountability of his past.

The notion of *estrangement* provides here a suggestive model for understanding Saer's stress on both the performative act of reading and of *writing travel* in *La ocasión*. Sveltana Boym observes, with remarks that are relevant to my analysis, that 'estrangement places emphasis on the process rather than the product of art [...] by making things strange, the artist does not simply displace them from an everyday context into an artistic framework; the artist also helps to "return sensation" to life itself'.[28] As I explained in the introduction to this chapter, *La ocasión* also calls upon 'returning sensation' to the performative character of literature.[29] Though the concept of estrangement has a long history, its relevance here lies in the Russian formalists' idea of 'defamiliarization' or *ostranenie*, whereby automatized perception is countered by literary devices that make writing, and reading, strange. As a literary concept, *ostranenie* was first theorized by Viktor Shklovsky in 1917, for whom art makes objects and forms 'unfamiliar' in order to increase perception. Connections were also established by critics between Shklovsky's formalist device of 'making strange' and another concept relevant to my reflections on defamiliarization: Bertolt Brecht's conception of *Verfremdung*, that is, 'making aware'. Although these terms differ

28 Sveltana Boym, 'Poetics and Politics of Estrangement: Victor Shklovsky and Hannah Arendt', *Poetics Today*, Vol. 26, 4 (2005), 581–611 (586).

29 It should be clarified, however, that the extent of the connection between Shklovsky's and Brecht's terms is a contested one, for despite links established by several scholars between those two notions, there are basic differences to them. As Peter Brooker argues, 'the idea that Brecht was directly indebted to Shklovsky [...] is unconvincing' for, compared to the extremely formal quality of Shklovsky's notion, 'Brecht's conception and use of *Verfremdung* [...] entailed a degree of political insight which thoroughly radicalised the formalist device of "making strange"', in Peter Brooker, 'Key Words in Brecht's Theory and Practice of Theatre', in *The Cambridge Companion to Brecht* (Cambridge: Cambridge University Press, 2006), 216.

in their specific coinages, both notions of 'making strange' and 'making aware' are also indicative of the removal of someone (or something) from its familiar surroundings and making them strange and thereby heightening their awareness of a place or situation. As such, both 'making strange' and 'making aware' resonate with the ways in which identity is coded in Saer's novel.[30] Propped up on mutability, the identity of the protagonist is always going through a process of constitution, always recently been and about to be. Thus, the comfort of closure is always denied, unsettling the reader in his search for concluding answers. In this regard, Laura García Moreno indicates that *La ocasión* gives us 'momentos en que la representación (aparentemente realista) entra en crisis y se desarma la idea de "evento" u ocasión como una unidad cerrada y claramente legible' [moments when representation (seemingly realist) is thrown into crisis and the idea of 'event' or occasion as a closed and clearly legible unit is upset].[31] Though García Moreno does not refer to the event as the performative act of reading but to the narrated events in the novel, her observations on the difficulties posed by Saer's novel for establishing both a realist and decisive reading of the events in the text tie in with my analysis of estrangement as a poetic that accentuates the performativity of reading, while reflecting the continual changes of the main character's identity through travel.

30 While I am indebted to the formulations on estrangement reviewed above, I depart specifically from the view of the formalists in their consideration of literature as a self-contained entity. Victor Shklovsky remarks in 'Art as Technique', that 'as a literary critic, I've been engaged in the study of the internal laws that govern literature. If I may bring up the analogy of a factory, then I would say that neither the current state of the world cotton market nor the politics of cotton trusts interests me. One thing alone concerns me: the number of strands that make up the cotton plant and the different ways of weaving them', in Víktor Shklovsky, *Theory of Prose*, trans. Benjamin Sher (Elmwood Park, IL: Dalkey Archive Press, 1990), vii. Concurring with the works of Said, Attridge and Eagleton examined in the introduction to this chapter I see the act of reading, not just as a recognizing of the 'weaving of strands' but as grounded in a worldly experience. In fact, in my view, the idea of estrangement as a pushing of the reader to an outside of the text, verifies my point.

31 Laura García Moreno, 'Entre la pampa del lenguaje y el alambrado de la escritura: la mancha roja como síntoma de la ambigüedad de lo real en *La ocasión* de Juan José Saer', *Journal of Iberian and Latin American Studies*, 23:2 (2017), 195–215 (201).

The desert

The analysis above on estrangement, as breaking the illusionism of the realist narrative and deepening the appreciation of reading as an event, seems to be at odds with the initial description of the desert offered in the novel. This space is conceived at first as subjected to Bianco's dominion and rationalization, an unoccupied region equated to a mental construct, characterized by its emptiness and lack of substance. But as the novel progresses, we see that that perception of the desert changes in a profound manner and comes to challenge the protagonist of *La ocasión* with an arresting materiality. Let us turn now to the initial portrayal of the desert offered in the novel:

> Le ha parecido, casi de inmediato, que por su monotonía silenciosa y desierta, la llanura era un lugar propicio a los pensamientos, no los rojizos y rugosos, del color de sus cabellos, como los que tiene ahora, sino sobre todo los pulidos, los incoloros que encastrándose unos en otros en construcciones inalterables y translúcidas, *le servirían para liberar a la especie humana de la servidumbre de la materia*. La extensión chata, sin accidentes, que lo rodea, gris como el cielo de finales de agosto, representa mejor que ningún otro lugar el vacío uniforme, el espacio despojado de la fosforescencia abigarrada que mandan los sentidos, la tierra de nadie transparente en el interior de la cabeza en la que silogismos estrictos y callados, claros, se concatenan.

> [it had appeared to him almost immediately, that because of its silent and deserted monotony, it was a place [the pampas] that favored reflection, not for the rough and resistant thoughts, of the same color as his hair, like those that he is now having, but for the smooth, colorless ones above all, those fitting together in unalterable and transparent structures, *which ought to serve him to liberate human space* [*sic*] *from the servitude of the matter.* The flat expanse, with no surface irregularities, that surrounds him, as grey as the sky at the end of August, represents better than any other place the uniform emptiness, the space devoid of phosphorescence of ill-assorted colors that the senses emit, that translucent no man's land inside his head where strict, silent and clear syllogisms concatenate.][32]

Against the precepts of positivism that Bianco wanted to refute, he sees the desert as favouring the construction of internal thoughts untainted by

32 Saer, *La ocasión*, 20–21, 2–3 (my emphasis).

the external world. Indeed, the passage insists on this feature: the desert appears to favour thoughts free of connections to meanings outside himself – rather than ones rooted in material reality (his hair, for instance). As if he was a transcendental phenomenologist, his thoughts and plans for future work represent a flight from matter. Against concrete experience and worldly events, Bianco at first sees the desert as a wordless mental platform, the very epitome of emptiness. The words that characterize both the deserted plains and the protagonist's state of mind when thinking underpin that feeling of vacuity: 'desierta', 'vacío uniforme', 'despojado', and 'la tierra de nadie transparente'. However, while at first these plains are depicted as empty where 'por leguas y leguas no se ve un árbol ni una piedra' [for leagues and leagues around, there is not a tree or a stone to be seen],[33] as the novel unfolds this depiction of the desert evolves until it becomes quite the contrary – a terrain where things and people prove to have a real materiality to themselves.

Although it is treated from different perspectives in each text, the contested emptiness of the desert is a feature that comes forth time and again in those of Saer's novels examined in this book. In his analysis of *La ocasión*, Julio Premat explains that representing the Pampas as an empty space is part of Saer's dialogue with textual and cultural traditions.[34] For Premat, just writing about the Pampas 'supone intervenir en una polémica que ha nacido y envejecido junto con la Nación que se fundó en su supuesta inmensidad' [means taking part in a controversy which was born and aged with the Nation which was founded in its alleged immensity].[35] In this regard, Alan Pauls points out that *La ocasión* carries out an 'ejercicio sutil de anacronismo' [a subtle exercise of anachronism] in which Saer alludes

33 Saer, *La ocasión*, 21, 3.

34 For another scholar, Isabel Quintana, Saer uses the yellow fever as a dramatic base to tell the story of Bianco. Isabel Quintana, 'La historia del error: segunda historia del error', in *Figuras de la experiencia en el fin de siglo, Cristina Peri Rossi, Ricardo Piglia, Juan José Saer, Silviano Santiago* (Rosario: Beatriz Viterbo, 2001), 127–137 (132).

35 Julio Premat, 'Pampas (*La Ocasión*)', in *La dicha de Saturno: Escritura y melancolía en la obra de Juan José Saer* (Rosario: Beatriz Viterbo, 2002), 320–344 (323) (translation mine).

to a repertoire of themes of the nineteenth century such as immigration and the colonization of the country-side, but they are seen from Bianco's gaze, which changes their constitution. Therefore, the elements from the nineteenth century are present in the novel but 'todo vibra de un modo extraño, con una luz nueva, como en el centro de una alucinación' [everything vibrates in a strange manner, with a new light, as in the centre of an hallucination].[36]

As in *Las nubes*, Saer also engages in *La ocasión* with travel narratives of the nineteenth century in Spanish America. Bianco's successful businesses, his exploitation of the land and his introduction of wire fencing to better administrate landed properties recall that body of literature written by travellers whom Mary Louise Pratt aptly terms 'the capitalist vanguard'. For Pratt, 'far from mystifying European expansionist designs [...] the capitalist vanguard tended to thematize them – indeed consecrate them'.[37] Pratt explains that these mercantile travellers, such as Francis Bond Head, John Miers, John Mawe and Gaspar Mollien replaced the previous aestheticizing and romanticizing rhetoric of discovery of previous explorers and naturalists, with a goal-oriented writing of achievement. Pratt sees the capitalist vanguard as constructing another facet of the 'reinvention of America', in the sense that these travellers encoded Latin American landscapes as terrain in need of the rationalized exploitation which could be brought about by the Europeans. For that reason, the landscape was deromanticized and 'primal nature held considerably less interest for these economic adventurers than it did for Humboldt'.[38] On the one hand, Bianco's views of the Pampas and his journey to Argentina tally with those of the capitalist vanguard in that they show a reduced interest in romanticizing and engaging aesthetically with the landscape in favour of a more utilitarian view of nature. On the other, there is a key factor which differentiates *La ocasión* from this approach. In many accounts by mercantile travellers 'the itinerary itself becomes the occasion for a narrative of success, in which travel is a

36 Alan Pauls, '*La ocasión*, mujer, gaucho malo, caballos', in *Zona de prólogos*, ed. Paulo Ricci (Buenos Aires: Seix Barral, 2011), 161–172 (168).

37 Mary Louise Pratt, *Imperial Eyes*, 144.

38 Pratt, *Imperial Eyes*, 146.

triumph in its own right'.[39] In *La ocasión*, in contrast, the outcome of travelling, though bringing about economic prosperity, is a far cry from being a 'triumph in its own right', for Bianco's chief intention, that is to refute positivism in that space that 'favours reflection', is in fact never realized.

In this regard, and in a similar note to that of Alan Pauls, Graciela Montaldo observes a conjunction of reality and mirage in *La ocasión*, particularly in its portrayal of the desert, which for her determines the ambiguity of this space. Montaldo observes that the plains themselves are thus simultaneously 'un exceso de realidad y un espejismo' [an excess of reality and a mirage].[40] She adds that, for the main character, the desert, Argentina, his own powers, his mother tongue, his origin, and his name are all an enigma to him. But part of that delusion, I believe, not only emerges from the fusion of elements from the nineteenth century with contemporary ones or the combination of reality and mirage, it also stems from Bianco's intention to withdraw into a speculative sphere as completely detached from his own circumstances. Notwithstanding the main character's intentions to understand the world and produce a body of work completely unaffected by anything outside his mental images: 'para hacer surgir de él, de sus inmediaciones desiertas y silenciosas, como golpes callados y gélidos, el pensamiento, en su doble expresión de puro y de pragmático' [so as to bring forth thought from him, from his deserted and silent surroundings, like ice-cold and muffled blows, in this dual expression as pure and as pragmatic reflections],[41] the desert and his journeys clash with those intentions as they prove to be undeniably caught up with Bianco's concrete material situation. In another echo of Cervantes's text, Saer's novel dramatizes the tension between reality and Bianco's will to flee into fantasy and delusion. The protagonist treats the desert as a spatial region that is there in its emptiness to allow for his thoughts to range, relying thus on nineteenth-century landscape conventions that naturalized and justified the dominion of the European settlers. But as the novel

39 Pratt, *Imperial Eyes*, 145.

40 Graciela Montaldo, 'La Pampa es lo más real', in *De pronto, el campo: literatura argentina y tradición rural* (Rosario: Beatriz Viterbo, 1993), 130–133 (131).

41 Saer, *La ocasión*, 93, 82.

progresses, the landscape proves not just not to be there to be used as an empty platform, it becomes the medium of the clash with reality and the means for reflections on the formation of identity. The following passage of the novel confirms this last point by introducing images of the relationship that Bianco has with the desert, not as one of dominion over the land but as one that is puzzling and problematic:

> En la llanura no se mueve nada, no se ve un pájaro, un animal, una nube, no sopla ninguna brisa, y el pasto beige que se apelmaza, de tan blando, bajo las botas de Bianco, no reluce en la luz irreal y cenicienta. 'Estoy soñando', piensa Bianco. 'Estoy sin duda en mi casa de París [...] y me he puesto a soñar, con imágenes despedazadas e incoherentes, que tuve una escaramuza con los positivistas, que me fui a Normandía, a Sicilia, que me hice adjudicar unos terrenos en la llanura, en el fin del mundo [...] y que ahora estoy en un espacio vacío, gris y beige, en el que ocurre nada, aparte del silencio de los sueños, del sueño de alguien que soy yo y que no sabe que está durmiendo en su cama, en un lugar que se llama París, que se llama mundo.
>
> [On the pampas nothing is moving, there is not a bird, an animal, a cloud in sight, there is no breeze blowing, and the tawny grass that is so soft that it collects in compact clumps beneath Bianco's boots has no sheen in the unreal, ashen light. 'I'm dreaming', Bianco thinks. 'I'm doubtless in my house in Paris, fast asleep [...] and I've begun to dream, with incoherent fragments of images, that I had a run-in with the positivists, that I went off to Normandy, to Sicily, that I secured a land grant on the pampas, at the end of the world [...] that I am now in an empty space, gray and beige, in which nothing is happening, apart from the silence characteristic of dreams, of the dream of someone who is myself and who does not know that he is sleeping in his bed, in a place called Paris, called the world'.][42]

In this extract the desert seems to be conveyed in a paraleptic manner, in the sense that the text denies by enumeration and the accumulation of negative adverbs precisely the features that should be affirmed about that space. We are reminded that, despite Bianco's intentions to free himself from matter, to be in a dream, he is not. He is, in fact, depicted as bound fast by his location in the very act of trying to set himself free from the concreteness of that space. As María del Pilar Blanco notes of some contemporary representations of the desert, this portrayal of Bianco in the

42 Saer, *La ocasión*, 190–191, 185.

desert poses, 'a troubling and negative "thick detail" that signals an altogether different idea of regenerative landscape'.[43] The images of the desert change around the protagonist at the same time as he begins to see that space as one of alienation rather than, as at the beginning of the novel, one of regeneration. I started this chapter explaining how the desert was intended by Bianco to be a space for retirement and productive reflection, but we observe here that the redemptive powers of the desert (or of Bianco in the desert) end up being altogether ineffective as the protagonist never comes up with the luminous pages he was supposed to produce there and is even profoundly confused as to whether he actually is on that terrain or in a dream. Thus, the desert becomes a medium of estrangement, not the empty landscape of the beginning of the novel that was to facilitate Bianco's thoughts, but a vehicle of perception and a trigger for the examination of the consequences of migration.

Let us now turn to how conceptions of identity are formulated in the novel in relation to Bianco's situation in the desert and his journeys:

> A fuerza de querer confundir al mundo en lo relativo a sus orígenes, está terminando por confundir él mismo sus orígenes, y lo que es opaco y brumoso para el mundo, ya lo es también para sí mismo, de modo que las máscaras sucesivas que ha ido llevando desde los comienzos inciertos, en un lugar incierto, ya no sabe bien cuál, las máscaras de La Valette, de Oriente, de Londres, de Prusia, de París, de Buenos Aires, se apelmazan, viscosas, contra su cara, y la deforman, la borran, la vuelven mera materia perecedera y residual, lo transforman a él mismo en el argumento viviente de los que odia, de los que arrancándole la máscara en París, creyendo descubrir su verdadera cara, dejaron en su lugar un agujero negro, que él va llenando, poco a poco, con títulos de propiedad, con ganado, con ese rancho en cuya puerta está ahora observando cómo Garay López, sacudiéndose sobre el caballo, se vuelve cada vez más diminuto en dirección al horizonte hasta que desaparece por completo.
>
> [By dint of wanting to confuse everyone as to his origins, he is ending up by confusing his origins himself, and what is opaque and hazy for everyone else is now that way for himself as well, so that the successive masks that he has been wearing from his uncertain beginnings, in an uncertain place – he no longer knows quite which – the

43 María del Pilar Blanco, *Ghost-Watching American Modernity: Haunting, Landscape, and the Hemispheric Imagination* (New York: Fordham University Press, 2012), 67.

> masks of Valletta, of the Orient, of London, of Prussia, of Paris, of Buenos Aires press, viscously, against his face, and deform it, efface it, turn it into mere perishable and residual material, transform his very own self into the living argument of those whom he detests, of those who ripping off his mask in Paris, thinking that they would discover his true face, left in its place a black hole, that he is little by little filling up with property rights, with cattle, with this cabin in whose doorway he is now observing how Garay López, posting up and down on his horse, is growing smaller and smaller on the horizon until he disappears completely.][44]

Bianco is initially lured into the trap of believing that identity is a voluntary construct. But later he realizes that what he thought were means of changing selves, his journeys and his fabricated stories, comprehend a changing of 'masks': a perpetual self-invention that has caused self-estrangement and left a void that he can only fill with objects. He cannot recognize who he is, just the objects that surround him. He cannot reconstruct a coherent representation of his own identity, but only the corporeality of what is around, which is, by its own material nature, earthborn and temporal. Bianco is unceasingly haunted by the past, which presses 'viscously, against his face' but it also 'effaces', transforms itself into a 'black hole' filled by objects that finally disappear. The result of the main character's permanent migrations, and the hauntedness of his past are thus evocative of what Edward Said and Theodor Adorno have called 'the principle of annulment in the life of émigrés'. For both scholars, 'the past life of émigrés is, as we know, annulled. [...] Anything that is not reified, cannot be counted and measured, ceases to exist'.[45] Bianco cannot understand his reality or comprehend his identity, nor can he prove positivism wrong. He tries to ease his estrangement from reality by being one with it and fails. He attempts to abridge the distancing that his journeys and experience of the desert seem to widen and also fails. His past realities, then, are annulled, since they cannot be 'reified', leaving only a void. Much of the narrative's tensions in *La ocasión* lie in Bianco's intention to withdraw from reality and conceive it as a voluntary mental

44 Saer, *La ocasión*, 93–94, 82–83.

45 Theodor Adorno, 'Reflections from the Damaged Life' in *Minima Moralia*, quoted in Edward Said, 'Between Worlds', *Reflections on Exile and Other Literary and Cultural Essays* (London: Granta Books, 2001), 562.

assemblage that denies both his connections and estrangement from it. Those tensions, which are also an effect of his journeys, relate to Álvaro Fernández Bravo's and Florencia Garramuño's reflections on migrancy:

> Nos gustaría reconocer al sujeto migrante como aquel que navega entre universos desiguales, sin obtener una certeza acerca de su identidad y más bien multiplicando las preguntas que lo acosan en el curso de su travesía. Si todas las identidades son múltiples, el migrante agudiza esa situación, a través de una afiliación dividida.
>
> [We would like to understand the migrant subject as that who navigates different universes, without obtaining any certainty about his identity, rather multiplying the questions that haunt him in his journeys. If all identities are multiple, the migrant exacerbates that situation through his divided affiliations.][46]

The drama that Bianco's thoughts produce, and which haunt him in his journeys, is another main theme in *La ocasión* and a further echo of Cervantes in terms of the nature of consciousness and the psychology of a character who travels back and forth from reality to delusion. This theme is present not only in the relationship that Bianco has with the desert and his understanding of the consequences of migration, but in all of his bonds with reality. For instance, from the moment when Bianco unexpectedly finds his wife with his business partner in a compromising situation (she is reclining on the sofa, pleasurably smoking a cigarette and he is sitting next to her, leaning forward, talking and smiling, in what Bianco deems to be a lascivious manner), we see him become emotionally obsessed with the idea that they had an affair. His obsession and extreme jealousy continue, and he becomes even more troubled soon after, when his wife announces to him that she is pregnant. He is devoured by his *ideé fixe* concerning the supposed infidelity and for almost all of the rest of the novel, we see him mired in his jealousy trying to either divine some proof of the affair or extract a confession. That encounter or *event* brings on a terrible, although

46 Álvaro Fernández Bravo, Florencia Garramuño 'Introducción', in *Sujetos en tránsito: (in)migración, exilio y diáspora en la cultura latinoamericana*, eds Álvaro Fernández Bravo, Florencia Garramuño and Saúl Sosnowski (Buenos Aires: Alianza, 2003), 11–28 (15) (my translation).

not quite exteriorized personal crisis. His thoughts take hold of him in such a manner that he never obtains deliverance from his jealousy, nor does he manage to clearly discern between reality and delusion. This is very clearly shown when the novel describes how the yellow fever hits the city of Buenos Aires. The main character is in such a blind pursuit of a confession from Garay López that he doesn't believe the people he comes across are ill. He prefers to think that, instead, his business partner and indeed the whole Garay López family have just had too much sun. It takes a couple of days and the spreading of the epidemic for him to see the reality: that the yellow fever has claimed them. At this point in the novel, the narrator pushes the reader out of the fiction again, making us reflect on the nature of the protagonist's blindness to see reality: 'se pierde […] las razones que han podido impulsar a Garay López a contarle la historia de la epidemia, sin que se le cruce una sola vez por la mente que lo que le ha dicho Garay López, lejos de ser una historia […] puede tener algún elemento de verdad' [he fruitlessly fritters away […] the reasons that might have impelled Garay López to tell him that story of the epidemic, without it ever once crossing his mind that what Garay López has told him, far from being a story […], might have a certain element of truth in it].[47] Beatriz Sarlo points out that despite his aspirations to telepathy, Bianco 'no puede ver nada, no puede entender ni a su mujer ni a su amigo, no puede predecir mínimamente el curso de las cosas' [cannot see a thing, he cannot understand his wife, his friend, he cannot even predict the course of events].[48]

La ocasión, therefore, is as much a novel about the experience of migration, the power that travel exercises over identity and perception of past experiences, as it is an attempt to fathom them in the reading as a literary event. Throughout this chapter I have intimated how part of the formal construction of Saer's novel distances the reader from the text, a distancing that also follows from the narrative of a character with obscure origins and the story of his migrations and resulting failures. In this way, the poetics of the novel support a narrative of distancing that occurs with the reading.

47 Saer, *La ocasión*, 183, 177.

48 Beatriz Sarlo, 'Mujer, pena y misterio', in *Escritos sobre literatura argentina* (Buenos Aires: Siglo XXI Editores Argentina, 2007), 281–285, 287.

Thus, at the same stroke, the experience of the text and in the text is one of distancing. This brings us neatly back to the discussion at the beginning of the chapter and indeed to the title of the novel, since distancing or estrangement, as Eagleton has pointed out, 'is also an event. It is language doing something to the reader, which is to say language as rhetoric'.[49] Explained in *La ocasión*'s own terms, this means that Saer's poetics in *La ocasión* are also events, both those taking place in the text and the event of literature for the reader.

49 Terry Eagleton, *The Event of Literature*, 190.

Conclusion

In his consideration of travel as metaphor, Georges Van Den Abbeele observes that 'in the view of this slippery path leading one back and forth between text and travel, [...] what might otherwise be construed as idle statements on travel in a writer's discourse allow on the contrary for the elaboration of a critical discourse of considerable force'.[1] Read together, and in the light of Van Den Abbeele's remark, the five texts discussed in this book make an effective contribution to this 'critical discourse' taking their force from the re-examination and re-elaboration of previous paradigms, escaping what could otherwise be 'idle statements' about culture. The journeys scrutinized in this book, while 'leading back and forth between text and travel', open up new perspectives from which to consider both travel and literature. Indeed, I would go further to say that, in the five novels considered here, it is the very textual machinery of the writing of fictional travel that facilitates the creation of different and alternative viewpoints. Additionally, in both Bolaño's and Saer's work, the journey concomitantly leads to an involvement with the design of spaces, highlighting further the critical potential of those designs. However, while Saer's and Bolaño's novels underscore the critical potency of travel and space, they do not simply celebrate movement, for they do not fall into the category of those discourses that commonly privilege 'distance and separation as aesthetic benefits while simultaneously deploring any political or psychological crises that such conditions may engender'.[2]

In *Writing Travel* I have shown how precedent travel narratives, which contributed to hegemonic views of Latin America and the nation, are questioned and revised in different ways. This is particularly apparent

1 Georges Van Den Abbeele, *Travel as Metaphor*, xxiii.

2 Caren Kaplan, *Questions of Travel: Postmodern Discourses of Displacement* (Durham, NC and London: Duke University Press, 1996), 94.

in the three novels by Saer, *Las nubes*, *El entenado* and *La ocasión*, and also in Bolaño's *Los detectives salvajes*. In this way, the texts I study here recognize the importance of previous travel narratives, and can be seen as echoing Jason Wilson's views (see Introduction), that the latter have, 'in complex ways created Latin America'.[3] However, Bolaño and Saer respond to that tradition through fictions written from the receiving end of factual travel writing in Latin America. Consequently, an objective of this study was to trace (in Chapters 1, 3 and 5), the ways in which Saer reprocesses, in turns, accounts of Conquest, foundational texts, and the writings of economically motivated travellers to Argentina; and does so while fashioning contemporary narratives of displacement and migration. The three novels examined in those chapters show Saer's concern with, and problematization of, history, travel and representation. Thus, *El entenado* exposes the flip-side of imperial discourses by highlighting failure within that discursive framework. *La ocasión* moves forward in time, grappling with inherited European discourses of travel as (financial) success in the nineteenth century, to articulate crises in negotiating identity and failure through migration. Finally, *Las nubes* has the most emphatic focus on the importance of travel writing in the constitution of national foundational fictions and while it relates to a national tradition, it does so in order to unsettle its basis by questioning the cogency of the national as a criteria and model for thinking about literature. On this last point, Saer responds to long-standing discussions on the validity of the national and to what has been, at times, presented as a requisite for literature written by Latin American writers.

The journeys of *Los detectives salvajes* take a similar course to that of *Las nubes*: high speed, the disconnections between the travellers and travelled territories, the cultivation of an aesthetic of fleeting looks and getting lost, all cast doubt on literature as deeply implicated with the national, and the nation as an explanatory literary form. In this way, the critical impulse of Bolaño's and Saer's texts coincides with Jean Franco's appraisal of Latin American novels of the second half of the twentieth century, which she

3 Jason Wilson, 'Travel literature', 803.

states make, 'visible that absence of any signified that could correspond to the nation' as such novels complicate the possibilities of reading them as any kind of national allegory, for the nation is seen as either 'a contested term or something like the Cheshire cat's grin – a mere reminder of a vanished body'.[4]

As I mentioned, *Writing Travel* is not only concerned with displacements, but also with the authors' treatment of space as a critical and related trope. Thus, that 'slippery path leading one back and forth between text and travel' relies, likewise, on the association of space and text. The space of the desert, for example, both in *2666* and in *La ocasión*, raises questions concerning the very role and nature of the literary discourse. In *2666* the desert acts as a platform to propose ways of reading which grant unfinished, and at times oscillating, meanings, that are actualized as they are 'un-hidden'. *La ocasión*'s spatial configurations foreground the problematic consequences of migration and invite the reader to reflect on estrangement, as both literary effect and the result of continuous displacements.

Finally, I have, in this book, sought to show how the deserts depicted in both Bolaño's and Saer's novels are not merely a static construct; they are constituted historically and have an important effect on both territories and people. In fact, we can consider the deserts depicted by both authors as contact zones, in the sense coined by Mary Louise Pratt, for without wanting to erase the geographical or historical specificities of the material deserts that inspired them, these deserts stage crossings, and are constitutional to cultural encounters and clashes. A key example of this is found in Saer's work as this author interrogates and reformulates the historical use of the desert as a cultural medium to naturalize exclusionary binaries – such as that of civilization and barbarism. Bolaño's *2666* is the text that most contemporizes the notion of the desert as contact zone, particularly through the intricate connections it establishes between cultural conflicts and drug trafficking, advanced capitalism, and the femicides. The critical potential of spatial practices and journeying

4 Jean Franco, 'The Nation as Imagined Community', in *The New Historicism*, ed. Aram Veeser (New York: Routledge, 1989), 204–212, 205.

is thus at the core of all the novels analysed in this study, with most of them casting a retrospective light which is then brought forward to the present. In *Writing Travel*, I have therefore sought to show how the writing of space and travel serves as a constitutive element and important critical lens through which to read the literary work of Roberto Bolaño and Juan José Saer.

Bibliography

Abbate, Florencia, 'La crisis de la identidad en el relato histórico: un análisis de *La ocasión* de Juan José Saer', *Amerika: Memoires, identities, territories*, 9 (2013), 78–89.

Adorno, Theodor W., *Minima Moralia: Reflections on a Damaged Life* (London and New York: Verso, 2005).

Adorno, Theodor W., *Notes to Literature*, Volume Two. Trans. Shierry Weber Nicholsen. Ed. Rolf Tiedemann (New York: Columbia University Press, 1992).

Ahmad, Aijaz , 'Jameson's Rhetoric of Otherness and the "National Allegory"', *Social Text*, 17 (1987), 3–25.

Aira, César, *Ema la cautiva* (Buenos Aires: Editorial de Belgrano, 1981).

Alweiss, Lilian, 'Heidegger and "The Concept of Time"', *History of the Human Sciences*, 15 (2002), 117–132.

Andermann, Jens, *Mapas de poder: una arqueología literaria del espacio argentino* (Rosario: Beatriz Viterbo, 2000).

Anderson, Benedict, *Imagined Communities: Reflections on the Origin and Spread of Nationalism*, revised edition (London and New York: Verso, 1991).

Anderson, Malcolm, *Frontiers: Territory and State Formation in the Modern World* (Cambridge: Polity Press, 1996).

Appadurai, Arjun, *Modernity at Large: Cultural Dimensions of Globalization* (Minneapolis: University of Minnesota Press, 1996).

Attridge, Derek, *J. M. Coetzee and the Ethics of Reading: Literature in The Event* (Chicago and London: The University of Chicago Press, 2004).

Attridge, Derek, *The Singularity of Literature* (London: Routledge, 2004).

Barclay, William, *The New Daily Study Bible: The Revelation of John*, Vol. 1 (Louisville, KY and London: Westminster John Knox Press, 2004).

Barclay, William, *The New Daily Study Bible: The Revelation of John*, Vol. 2 (Louisville, KY and London: Westminster John Knox Press, 2004).

Baudelaire, Charles, *Les fleurs du mal* (Paris: Gallimard, 2004).

Bhabha, Homi K., 'The Commitment to Theory', in *The Location of Culture* (New York: Routledge, 2008), 28–56.

Bhabha, Homi K., 'Dissemination: Time, Narrative and the Margins of the Modern Nation', in *The Location of Culture* (New York: Routledge, 2008), 199–244.

Birns, Nicholas, and Juan E. De Castro, *Roberto Bolaño as World Literature* (New York and London: Bloomsbury Academic, 2017).
Blanco, María del Pilar, *Ghost-Watching American Modernity: Haunting, Landscape, and the Hemispheric Imagination* (New York: Fordham University Press, 2012).
Bolaño, Roberto, *2666* (Barcelona: Anagrama, 2004).
Bolaño, Roberto, *2666*, trans. Natasha Wimmer (London: Picador, 2009).
Bolaño, Roberto, 'Acerca de *Los detectives salvajes*', in *Entre paréntesis* (Barcelona: Anagrama, 2006), 326–327.
Bolaño, Roberto, *Amuleto* (Barcelona: Anagrama, 1999).
Bolaño, Roberto, 'De aquí a la eternidad: interview with Sergio Paz', *El mercurio*, 13 (1998), 4–6.
Bolaño, Roberto, *El gaucho insufrible* (Barcelona: Anagrama, 2003).
Bolaño, Roberto, *Estrella distante* (Barcelona: Anagrama, 1996).
Bolaño, Roberto, 'Exilios' in *Entre paréntesis* (Barcelona: Anagrama, 2004), 49–59.
Bolaño, Roberto, 'Final: "*Estrella distante*" (Entrevista de Mónica Maristain)', in *Entre paréntesis* (Barcelona: Angarma, 2004), 329–344.
Bolaño, Roberto, 'Interview with Ana Anabitarte', *El Universal*, 5 July 1999.
Bolaño, Roberto, 'Interview with Gonzalo Núñez, 'Soy un pobre aprendiz', *La hora*, 3 November 1998, 30.
Bolaño, Roberto, 'Literatura y exilio', in *Entre paréntesis* (Barcelona: Anagrama, 2004), 40–46.
Bolaño, Roberto, *Los detectives salvajes* (Barcelona: Anagrama, 1998).
Bolaño, Roberto, *Nocturno de Chile* (Barcelona: Anagrama, 2000).
Bolaño, Roberto, 'Prólogo a las aventuras de Huckleberry Finn', in Mark Twain's, *Las aventuras de Huckleberry Finn*, trans. J. A. Larrinaga (Barcelona: Random House Mondadori, 2006).
Bolaño, Roberto, 'Sergio González Rodríguez bajo el huracán', in *Entre paréntesis* (Barcelona: Anagrama, 2004), 214–216.
Bolognese, Chiara, *Pistas de un naufragio: cartografía de Roberto Bolaño* (Santiago de Chile: Margen, 2009).
Borges, Jorge L., 'El escritor argentino y la tradición', in *Discusión*, in *Obras completas* (Buenos Aires: Emecé, 1969), 151–162.
Borges, Jorge L., 'Historia del guerrero y la cautiva', in *El aleph*, 1st edition 1949 (Buenos Aires: Emecé, 1957), 47–52.
Borm, Jan, 'Defining Travel: On the Travel Book, Travel Writing and Terminology', in *Perspectives on Travel Writing*, eds Glenn Hopper and Tim Youngs (Aldershot: Ashgate, 2004), 13–26.
Boym, Sveltana, 'Poetics and Politics of Estrangement: Victor Shlovsky and Hannah Arendt', *Poetics Today*, Vol. 26., 4 (2005), 581–611 (586).

Brizuela, Leopoldo, *El placer de la cautiva* (Buenos Aires: Temas, 2001).
Brooker, Peter, 'Key Words in Brecht's Theory and Practice of Theatre', in *The Cambridge Companion to Brecht* (Cambridge: Cambridge University Press, 2006).
Bull, Malcolm, 'On Making Ends Meet', in *Apocalypse Theory: And the Ends of the World*, ed. Malcolm Bull (Oxford: Wiley-Blackwell, 1995), 1–20.
Burgos, Carlos, 'Roberto Bolaño: La violencia, el mal, la memoria', *Nuevo texto crítico*, Vol. XXII, 43–44 (2009), 123–144.
Busaniche, José Luis, *Historia argentina* (Buenos Aires: Solar-Hachette, 1973).
Canaparo, Claudio, *Geo-epistemology: Latin America and the Location of Knowledge* (Oxford: Peter Lang, 2009).
Cerbone, David R., *Heidegger: A Guide for the Perplexed* (London and New York: Continuum International Publishing Group, 2008).
Chambers, Iain, *Migrancy, Culture, Identity* (London, Routledge, 1994).
Chanady, Amaryll, 'Saer's Fictional Representation of the Amerindian Context of Modern Historiography', in *Amerindian Images and the Legacy of Colombus,* eds René Jara and Nicholas Spadaccini (Minneapolis: University of Minnesota Press, 1992), 678–708.
Chiaramonte, Carlos, *Ciudades, provincias, estados: orígenes de la nación argentina 1800–1846* (Buenos Aires: Emecé, 2007).
Chiaramonte, Carlos, *Nación y estado en Iberoamérica: el lenguaje político en tiempos de las independencias* (Buenos Aires: Sudamericana, 2004).
Colombi, Beatriz, *Viaje intelectual: migraciones y desplazamientos en América Latina (1880–1915)* (Rosario: Beatriz Viterbo, 2004).
Colón, Cristobal, *Cartas que escribió sobre el descubrimiento de América y testamento que hizo a su muerte* (Madrid: Biblioteca Universal, 1880).
De Grandis, Rita, 'The First Colonial Encounter in *El entenado* by Juan José Saer: Paratextuality and History in Postmodern Fiction', *Latin American Literary Review*, 21 (1993), 30–38.
De las Casas, Bartolomé, *Brevísima relación de la destrucción de las Indias* (Madrid: Biblioteca de la historia, 1985).
De los Ríos, Valeria, 'Mapas y fotografías en la obra de Roberto Bolaño', in *Bolaño salvaje*, eds Edmundo Paz Soldán and Gustavo Faverón Patriau (Barcelona: Candaya, 2008), 237–258.
Deckard, Sharae, 'Peripheral Realism, Millenial Capitalism, and Roberto Bolaño's *2666*', *Modern Language Quaterly*, Vol. 3, 3 (2012), 351–372.
Derbyshire, Philip, '*Los Detectives Salvajes*: Line, Loss and the Political', *Journal of Latin American Cultural Studies: Travesia*, Vol. 18, 2–3 (2009), 167–176.
Derrida, Jacques, 'The Law of Genre', *Glyph 7* (1980), 202–232.

Derrida, Jacques, *Specters of Marx*, trans. Peggy Kamuf (New York and London: Routledge, 1994).

Dettelbach, Cynthia, *In the Driver's Seat: The Automobile in American Literature and Popular Culture* (Westport, CT: Greenwood Press, 1976).

Díaz-Quiñones, Arcadio, '*El entenado:* las palabras de la tribu', *Hispamérica*, XXI: 63 (1992), 3–14.

Domínguez Michael, Christopher, 'Bolaño, Roberto (Santiago de Chile 1953, Barcelona, España – 2003)', *Anales de literatura chilena* (2009), 251–256 (252).

Donoso Macaya, Ángeles, 'Estética, política y el posbile territorio de la ficción en *2666* de Roberto Bolaño', *Revista hispánica moderna*, vol. 62 (2009), 125–142.

Duffy, Edna, *The Speed Handbook: Velocity, Pleasure and Modernism* (Durham, NC: Duke University Press, 2009).

Eagleton, Terry, 'A conversation with Terry Eagleton', interview with José Manuel Barbeito Varela, *Atlantis*, vol. 23, nº 2 (2001), 169–185, 176.

Eagleton, Terry, *The Event of Literature* (New Haven, CT and London: Yale University Press, 2012).

Eagleton, Terry, *Literary Theory: An Introduction* (Oxford: Blackwell Publishers, 1983).

Eagleton, Terry, *On Evil* (New Haven, CT and London: Yale University Press, 2010).

Echevarría, Ignacio, 'Nota a la primera edición', in Roberto Bolaño's *2666* (Barcelona: Anagrama, 2004), 1123.

Echeverría, Esteban, *La cautiva* (Madrid: Cátedra, 1986).

Elsner, Jaś, and Joan-Pau Rubiés, *Voyages and Visions: Towards a Cultural History of Travel* (London: Reaktion Books, 1999).

Espinosa, Patricia, 'Estudio Preliminar', in *Territorios en fuga*, ed. Patricia Espinosa (Santiago: Frasis Editores Colección Ensayos, 2003), 13–32.

Espinosa, Patricia, 'Roberto Bolaño: un territorio por armar', in *Roberto Bolaño: la escritura como tauromaquia*, ed. Celina Manzoni (Buenos Aires: Corregidor, 2006), 125–132.

Espinosa, Patricia, 'Secreto y simulacro en *2666* de Roberto Bolaño', *Estudios filológicos*, 41 (2006), 71–79.

Farred, Grant, 'The Impossible Closing: Death, Neoliberalism, and the Postcolonial in Bolaño's *2666*', *Modern Fiction Studies*, Vol. 56, 4 (2010), 689–708.

Fass Emery, Amy, 'Sa(l)vage Ethnography: The Cannibalistic Imagination in Juan José Saer's *El entenado*', in *The Anthropological Imagination in Latin American Literature* (Columbia: University of Missouri Press, 1996), 109–125.

Fernández Bravo, Álvaro, and Florencia Garramuño et al., *Sujetos en tránsito: (in)migración, exilio y diáspora en la cultura latinoamericana*, eds Álvaro Fernández Bravo, Florencia Garramuño and Saúl Sosnowski (Buenos Aires: Alianza, 2003).

Franco, Jean, *Cruel Modernity* (Durham, NC: Duke University Press, 2013).
Franco, Jean, *The Decline and Fall of the Lettered City: Latin America in the Cold War* (Cambridge, MA and London: Harvard University Press, 2002).
Franco, Jean, 'Literature and Nationalism', in *An Introduction to Spanish-American Literature* (London: Cambridge University Press, 1969), 46–74.
Franco, Jean, 'The Nation as Imagined Community', in *The New Historicism*, ed. Aram Veeser (New York: Routledge, 1989), 204–212.
Franco, Jean, 'Pastiche in Contemporary Latin American Literature', in *Critical Passions: Selected Essays* (Durham, NC and London: Duke University Press, 1999), 393–404.
Franco, Jean, 'Questions for Bolaño', *Journal of Latin American Cultural Studies*, 18:2 (2009), 207–217.
Frantz, Julien, 'Prosopopée pour anapocalypse', *Ciclocosmia: Roberto Bolaño*, III (2010), 67–73.
Foucault, Michel, *Les mots et les choses* (Paris: Gallimard, 1966).
Foucault, Michel, *Madness and Civilization: A History of Insanity in the Age of Reason* (New York: Vintage Books, 1988).
Fourez, Cathy, '*2666* de Roberto Bolaño: los basureros de Santa Teresa, territorios del tiempo del fin', in *Los imaginarios apocalípticos en la literatura hispanoamericana contemporánea* (Bern: Peter Lang, 2010), 231–244.
Galdo, Juan Carlos, 'Fronteras del mal/ genealogías del horror: *2666* de Roberto Bolaño', *Hipertexto*, 2 (2005) <http://www.utpa.edu/dept/modlang/hipertexto/docs/Hiper2Galdo.pdf> [accessed 6 September 2008].
García Canclini, Néstor, *Imagined Globalization*, trans. George Yúdice (Durham, NC and London: Duke University Press, 2014).
García Moreno, Laura, 'Entre la pampa del lenguaje y el alambrado de la escritura: la mancha roja como síntoma de ambigüedad de lo real en *La ocasión* de Juan José Saer', *Journal of Iberian and Latin American Studies*, 23:2 (2017), 195–215.
Garramuño, Florencia, *Genealogías culturales, Argentina, Brasil y Uruguay en la novela contemporánea (1981–1991)* (Rosario: Beatriz Viterbo, 1997).
Garramuño, Florencia, *La experiencia opaca* (Buenos Aires: Fondo de cultura económica, 2009).
Garramuño, Florencia, 'Las ruinas y el fragmento. Experiencia y narración en *El entenado* y *Glosa*', in *Glosa, El entenado*, critical edn by Julio Premat (Poitiers: Université de Poitiers, Centre de recherches latino-américaines-Archivos; Córdoba: Alción editora, 2010), 710–728.
Gilman, Claudia, and Graciela Montaldo, *Preciosas cautivas* (Buenos Aires: Alfaguara, 1993).

Gollnick, Brian, 'El color justo de la patria: agencias discursivas en *El entenado* de Juan José Saer', *Revista de crítica literaria latinoamericana*, 57 (2003), 107–124.

Golomb Dettelbach, Cynthia, *In the Driver's Seat: The Automobile in American Literature and Popular Culture* (Westport, CT: Greenwood Press, 1976)

Gramuglio, María Teresa, 'El lugar de Juan José Saer', in *Ficciones argentinas: antología de lecturas críticas* (Buenos Aires: Norma, 2004), 329–366.

Greenblatt, Stephen, *Marvellous Possessions: The Wonder of the New World* (Oxford: Clarendon Press, 1991).

Hall, Stuart, 'Cultural Identity and Diaspora', in *Identity: Community, Culture, Difference*, ed. J. Rutherford (London: Lawrence & Wishart, 1990), 222–237.

Halperín Donghi, Tulio, *Proyecto y construcción de una nación, 1846–1880* (Caracas: Bibliotecha Ayacucho, 1980).

Halperín Donghi, Tulio, *Revolución y guerra: Formación de una elite dirigente en la Argentina criolla* (Buenos Aires: Siglo Veintiuno, 2005).

Halperín Donghi, Tulio, 'Sarmiento y el historicismo romántico' (Alicante: Biblioteca virtual Miguel de Cervantes, 2011) <http//www.cervantesvirtual.com/nd/ark:/59851/bmcws9b0> [accessed 12 December 2016].

Hartwig, Susanne, 'Jugar al detective: el desafío de Roberto Bolaño', *Iberoamericana*, 28 (2007), 53–71.

Heidegger, Martin, *The Essence of Truth: On Plato's Cave Allegory and Theaetetus* (London: Continuum, 2009).

Hobsbawm, Eric, *Nations and Nationalism since 1780: Programme, Myth, Reality* (Cambridge: Cambridge University Press, 1990).

Holland, Patrick, and Graham Huggan, *Tourists with Typewriters: Critical Reflections on Contemporary Travel Writing* (Ann Arbor: University of Michigan Press, 2000).

Iparaguirre, Sylvia, *La tierra del fuego* (Madrid: Suma de letras, 2001).

Jameson, Fredric, *The Political Unconscious: Narrative as a Socially Symbolic Act* (London and New York: Routledge, 2002).

Jameson, Fredric, 'Progress versus Utopia: Or, Can We Imagine the Future?', *Science Fiction Studies*, Vol. 9, N. 2 (1982), 147–158.

Jameson, Fredric, 'Third-World Literature in the Era of Multinational Capitalism', *Social Text*, 15 (1986), 65–88.

Kaplan, Caren, *Questions of Travel: Postmodern Discourses of Displacement* (Durham, NC and London: Duke University Press, 1996).

Kermode, Frank, *The Sense of an Ending: Studies in the History of Fiction* (Oxford: Oxford University Press, 2000).

Kohan, Martín, 'Historia y literatura: la verdad de la narración', in *Historia crítica de la literatura argentina: la narración gana la partida*, vol. 11, ed. Elsa Drucaroff (Buenos Aires: Emecé, 2000), 245–257.

Kohan, Martín, *Los cautivos* (Buenos Aires: Sudamericana, 2000).
Kohan, Martín, 'Saer, Walsh: una discusión política en la literatura', *Nuevo texto crítico*, 12–13 (1993–1994), 121–130.
Kraniauskas, John, 'Border Issues', *Travesia* 3 1–2 (1994), 5–13.
Kraniauskas, John, 'A Monument to the Unknown Worker: Roberto Bolaño's *2666*', *Radical Philosophy* 200 (2016), 37–46.
Kristeva, Julia, *Life: Hannah Arendt or Action as Birth and Estrangement*, vol. I of *Female Genius*, trans. Ross Guberman (New York: Columbia University Press, 2000).
Kunz, Marco, 'Cuando Bolaño "habla" mexicano, en torno a la oralidad fingida en *Los detectives salvajes*', in *Roberto Bolaño. Estrella cercana. Ensayos sobre su obra*, eds Augusta López Bernasocchi and José Manuel López de Abiada (Madrid: Verbum, 2012), 148–175.
Lawrence, D. H., *Apocalypse* (London: Penguin, 1980).
Leed, Eric J., *The Mind of the Traveller: From Gilgamesh to Global Tourism* (New York: Basic Books, 1991).
Lefebvre, Herni, *The Production of Space* (Oxford: Blackwell, 1991).
Levinson, Brett, 'Case Closed: Madness and Dissociation in *2666*', *Journal of Latin American Cultural Studies*, 18 (2009), 177–191.
Lindsay, Claire, *Contemporary Travel Writing of Latin America* (New York: Routledge, 2010).
Link, Daniel, 'Medi(t)acciones de lo real en *El entenado*', in *La Chancha con cadenas: doce ensayos de literatura argentina* (Buenos Aires: Ediciones del eclipse, 1994).
Lisle, Debbie, 'Between Fact and Fiction: The Generic Boundaries of Travel Writing', in *The Global Politics of Contemporary Travel Writing* (Cambridge: Cambridge University Press, 2006).
Livon-Grosman, Ernesto, *Geografías imaginarias: el relato de viaje y la construcción del espacio patagónico* (Rosario: Beatriz Viterbo, 2003).
Lopez, Kimberle S., 'Loving Cannibalism: Cannibalism and Colonial Desire in Juan José Saer's *El entenado*', in *Latin American Novels of the Conquest: Reinventing the New World* (Columbia and London: University of Missouri Press, 2002), 28–70.
López-Vicuña, Ignacio, 'The Violence of Writing: Literature and Discontent in Roberto Bolaño's "Chilean" Novels', *Journal of Latin American Cultural Studies*, 18:2 (2009), 155–166.
Ludmer, Josefina, *Aquí América Latina: una especulación* (Buenos Aires: Eterna Cadencia, 2010).
McGinn, Bernard, 'Revelation', in *The Literary Guide to the Bible*, eds Robert Alter and Frank Kermode (Cambridge, MA: The Belknap Press of Harvard University Press, 1990), 523–544.

Manzi, Joaquín, 'El secreto de la vida (no está en los libros)', in *Roberto Bolaño: la escritura como tauromaquia*, ed. Celina Manzoni (Buenos Aires: Corregidor, 2006), 153–167.

Manzoni, Celina, 'Biografías mínimas/ínfimas y el equívoco del mal', in *Roberto Bolaño: la escritura como tauromaquia*, ed. Celina Manzoni (Buenos Aires: Corregidor, 2006), 17–32.

Manzoni, Celina, 'Ficción de futuro y lucha por el canon en la narrativa de Roberto Bolaño', in *Jornadas homenaje a Roberto Bolaño (1953–2003) Simposio Internacional*, ed. Ramón González Férriz (Barcelona: ICCI Casa América a Catalunya, 2005), 27–47.

Manzoni, Celina, 'Narrar lo inefable. El juego del doble y los desplazamientos en *Estrella distante*', in *Roberto Bolaño: la escritura como tauromaquia*, ed. Celina Manzoni (Buenos Aires: Corregidor, 2006), 39–50.

Manzoni, Celina, 'Rara avis: *Nocturno de Chile*', in *La memoria de la dictadura: 'Nocturno de Chile' de Roberto Bolaño, 'Interrupciones 2' de Juan Gelman*, ed. Fernando Moreno (Paris: Ellipses, 2006), 49–55.

Manzoni, Celina, 'Recorridos urbanos, fantasmagoría y espejismo en *Amuleto*', in *Roberto Bolaño, una literatura infinita*, ed. Fernando Moreno (Poitiers: Centre de Recherches Latino-américaines/Archivos, 2005), 173–186.

Manzoni, Celina, 'Reescritura como desplazamiento y anagnórisis en *Amuleto*', in *Roberto Bolaño: la escritura como tauromaquia*, ed. Celina Manzoni (Buenos Aires: Corregidor, 2006), 175–184.

Masoliver Ródenas, Juan Antonio, 'Espectros mexicanos', in *Roberto Bolaño: la escritura como tauromaquia*, ed. Celina Manzoni (Buenos Aires: Corregidor, 2006), 65–69.

Mazzota, Giuseppe, 'The Emergence of Modernity and the New World', in *New Worlds and the Italian Renaissance: Contributions to the History of European Intellectual Culture* (Leiden and Boston, MA: Brill, 2012), 9–24.

Merrim, Stephanie, 'The First Fifty Years of Hispanic New World Historiography the Caribbean, Mexico, and Central America', in *The Cambridge History of Latin American Literature*, eds Roberto González Echevarría and Enrique Pupo Walker, Vol. I (Cambridge: Cambridge University Press, 1996), 58–100.

Meyer Spacks, Patricia, *Boredom: The Literary History of a State of Mind* (Chicago and London: The University of Chicago Press, 1995).

Migden Socolow, Susan, 'Spanish Captives in Indian Societies: Cultural Contact along the Argentine Frontier, 1600–1835', *The Hispanic American Historical Review*, 71, 1 (1992), 73–99.

Miller, Elaine, *Julia Kristeva on Philosophy and Art in Depressed Times* (New York: Columbia University Press, 2014).

Miller, Nicola, *In the Shadow of the State: Intellectuals and the Quest for National Identity in Twentieth-Century Spanish America* (London: Verso, 1999).

Montaldo, Graciela, *Ficciones culturales y fábulas de identidad en América Latina* (Rosario: Beatriz Viterbo Editora, 1999).

Montaldo, Graciela, *Juan José Saer: 'El limonero real'* (Buenos Aires: Biblioteca crítica Hachette, 1986).

Montaldo, Graciela, 'La Pampa es lo más real', in *De pronto, el campo: literatura argentina y tradición rural* (Rosario: Beatriz Viterbo, 1993), 130–133.

Moreiras, Alberto, 'Infrapolitics and the Thriller: A Prolegomenon to Every Possible Form of Antimoralist Literary Criticism. On Héctor Aguilar Camín's *La guerra de Galio* and *Morir en el Golfo*', in *The Ethics of Latin American Literary Criticism: Reading Otherwise* (New York: Palgrave Macmillan, 2007), 147–179.

Moretti, Franco, *Signs Taken for Wonders: On the Sociology of the Literary Forms* (New York and London: Verso, 2005).

Mounce, Robert H., *The Book of Revelation* (Grand Rapids, MI: W. B. Eerdmans, 1998).

Newman, Kim, *Wild West Movies, or How the West Was Found, Won, Lost, Lied About, Filmed and Forgotten* (London: Bloomsbury, 1990).

Núñez Cabeza de Vaca, Álvar, *Naufragios* (Madrid: Aguilar, 1945).

Operé, Fernando, *Historias de la frontera: el cautiverio en la América hispánica* (Buenos Aires: Fondo de Cultura Económica, 2001).

Paes de Barros, Deborah, 'Driving that Highway to Consciousness: Late Twentieth-Century American Travel Literature', in *The Cambridge Companion to American Travel Writing*, eds Alfred Bendixen and Judith Hamera (Cambridge: Cambridge University Press, 2009), 228–243

Pagden, Anthony, *European Encounters with the New World* (London: Yale University Press, 1994).

Parkinson Zamora, Lois, *The Usable Past: The Imagination of History in Recent Fiction of the Americas* (Cambridge: Cambridge University Press, 1997).

Parkinson Zamora, Lois, *Writing the Apocalypse: Historical Vision in Contemporary U. S. and Latin American Fiction* (Cambridge: Cambridge University Press, 1989).

Pastor, Beatriz, *Discursos narrativos de la conquista: mitificación y emergencia* (Hanover: Ediciones del Norte, 1988).

Patruno, Luigi, *Relatos de regreso: ensayos sobre la obra de Juan José Saer* (Rosario: Beatriz Viterbo, 2015).

Pauls, Alan, '*La ocasión*, mujer, gaucho malo, caballos', in *Zona de prólogos*, ed. Paulo Ricci (Buenos Aires: Seix Barral, 2011), 161–172.

Paz Soldán, Edmundo, 'Introducción, Roberto Bolaño: literatura y Apocalipsis', *Bolaño salvaje*, eds Edmundo Paz Soldán and Gustavo Faverón Patriau (Barcelona: Candaya, 2008), 11–30.

Poblete Alday, Patricia, *Bolaño: otra vuelta de tuerca* (Santiago de Chile: Universidad academia de humanismo cristiano, 2010).
Pons, María Crisitina, *Memorias del olvido: Del Paso, García Márquez, Saer y la novela histórica de fines del siglo XX* (Mexico: Siglo XXI, 1996).
Porter, Dennis, *Haunted Journeys: Desire and Transgression in European Travel Writing* (Princeton, NJ and Oxford: Princeton University Press, 1991).
Pratt, Mary Louise, *Imperial Eyes: Travel Writing and Transculturation* (New York: Routledge, 2008).
Premat, Julio, 'El eslabón perdido: *El entenado* en la obra de Juan José Saer', *Caravelle*, 66 (1996), 75–93.
Premat, Julio, 'Mitos (*El limonero real, La ocasión, La pesquisa, Las nubes*)', in *La dicha de Saturno: escritura y melancolía en la obra de Juan José Saer* (Rosario: Beatriz Viterbo, 2002), 330–344.
Premat, Julio, 'Pampas (*La Ocasión*)', in *La dicha de Saturno: escritura y melancolía en la obra de Juan José Saer* (Rosario: Beatriz Viterbo, 2002), 320–344.
Prieto, Adolfo, *Los viajeros ingleses y la emergencia de la literatura argentina 1820–1850* (Buenos Aires: Sudamericana, 1996).
Primeau, Ronald, *Romance of the Road: The Literature of the American Highway* (Bowling Green, OH: Bowling Green State University Popular Press, 1996).
Quintana, Isabel, '*Figuras de la experiencia en el fin de siglo, Cristina Peri Rossi, Ricardo Piglia, Juan José Saer, Silviano Santiago* (Rosario: Beatriz Viterbo, 2001).
Reati, Fernando, '*Las nubes* de Juan José Saer: Un viaje por la pampa hacia otra metafísica de lo real', *Inti*, 52–53 (2000–2001), 281–294.
Ricoeur, Paul, *Freud and Philosophy: An Essay on Interpretation* (New Haven, CT and London: Yale University Press, 1970).
Riera, Gabriel, 'La ficción de Saer: ¿una "antropología especulativa"? (una lectura del *El entenado*)', *MLN*, 111.2 (1996), 368–390.
Rivas, Luis H., *¿Qué dice el Apocalipsis?* (Buenos Aires: Pan y Trabajo, 1987).
Romano Thuesen, Evelia, '*El entenado*: relación contemporánea de las memorias de Francisco Del Puerto', *Latin American Literary Review*, 23 (1995), 43–63.
Ross, Kathleen, 'Historians of the Conquest and Colonization of the New World 1550–1620', in *The Cambridge History of Latin American Literature*, eds Roberto González Echevarría and Enrique Pupo Walker, Vol. I (Cambridge: Cambridge University Press, 1996), 101–142.
Rotker, Susana, *Captive Women: Oblivion and Memory in Argentina*, trans. Jennifer French (Minneapolis and London: University of Minnesota Press, 2002).
Rowe, William, 'The Place of Literature in Cultural Studies', in *Contemporary Latin American Cultural Studies*, eds Stephen Hart and Richard Young (London: Arnold, 2003), 37–47.

Rubiés, Joan-Pau, 'Futility in the New World: Narratives of Travel in Sixteenth-Century America' in *Voyages & Visions: Towards a Cultural History of Travel*, eds Jás Elsner and Joan-Pau Rubiés (London: Reaktion Books Ltd, 1999), 74–100.
Saccomanno, Guillermo, *La lengua del malón* (Buenos Aires: Planeta, 2003).
Saer, Juan José, 'Dos razones', in *La narración-objeto* (Buenos Aires: Seix Barral, 1999), 155–166.
Saer, Juan José, 'Ebelot', in *El concepto de ficción* (Buenos Aires: Ariel, 1997), 71–78.
Saer, Juan José, *El entenado* (Habana: Fondo Editorial Casa de las Américas, 2006)
Saer, Juan José, 'El escritor argentino en su tradición', in Saer's *Trabajos* (Buenos Aires: Seix Barral, 2006), 61–67.
Saer, Juan José, 'El valor del mito' in *La historia y la política en la ficción argentina* (Santa Fé: Centro de Publicaciones Universidad Nacional del Litoral, 1995), 73–86.
Saer, Juan José, *The Event*, trans. Helen Lane (London: Serpent's Tail, 1995).
Saer, Juan José, *La ocasión* (Barcelona: Círculo de lectores, 1988).
Saer, Juan José, 'La perspectiva exterior, Grombowicz en Argentina', in *El concepto de ficción* (Buenos Aires: Seix Barral, 2010), 17–29.
Saer, Juan José, *Las nubes* (Buenos Aires: Seix Barral, 1997).
Saer, Juan José, 'Líneas del Quijote', in *La narración objeto* (Buenos Aires: Seix Barral, 1999), 31–54.
Saer, Juan José, 'Literatura y crisis argentina', in *El concepto de ficción* (Buenos Aires: Seix Barral, 2010), 94–120.
Saer, Juan José, 'Memoria del río', *Clarín*, 27 February (2000) <http://www.clarin.com/suplementos/cultura/2000/02/27/e-00501d.htm> [accessed 6 March 2009].
Saer, Juan José, 'Nuevas deudas con el *Quijote*', in *Trabajos* (Buenos Aires: Seix Barral, 2006), 79–82.
Said, Edward, *Orientalism* (London: Penguin, 2003).
Said, Edward, *Reflections on Exile and Other Literary and Cultural Essays* (London: Granta Books, 2001).
Said, Edward, *The World, The Text, and The Critic* (Cambridge, MA: Harvard University Press, 1983).
Sarlo, Beatriz, 'Aventuras de un médico filósofo', in *Escritos sobre literatura argentina* (Buenos Aires: Siglo XXI editores, 2007), 296–303.
Sarlo, Beatriz, 'De la voz al recuerdo', in *Escritos sobre literatura argentina* (Buenos Aires: Siglo XXI Editores, 2007), 312–316.
Sarlo, Beatriz, 'En el origen de la cultura argentina: Europa y el desierto', in *Escritos sobre literatura argentina* (Buenos Aires: Siglo Veintiuno Editores, 2007), 25–29.

Sarlo, Beatriz, 'La política, la devastación', in *Glosa, El entenado*, critical edn by Julio Premat (Poitiers: Université de Poitiers, Centre de recherches latino-américaines-Archivos; Córdoba: Alción editora, 2010), 762–778.

Sarlo, Beatriz, 'Mujer, pena y misterio', *Escritos sobre literatura argentina* (Buenos Aires: Siglo XXI editores, 2007), 286–288.

Sarlo, Beatriz, *Tiempo presente: notas sobre el cambio de una cultura* (Buenos Aires: Siglo veintiuno editores, 2001).

Sarmiento, Domingo F., *Facundo o civilización y barbarie* (Buenos Aires: Sopena, 1960).

Saucedo Lastra, Fernando, 'El país enemigo: México en la obra de Roberto Bolaño 1980–2004' (unpublished doctoral thesis, Universitat Pompeu Fabra, 2012), 108. <http:hdl.handle.net/10230/19805> [accessed 4 June 2013].

Sauri, Emilio, '"A la pinche modernidad": Literary Form and the End of History in Roberto Bolaño's *Los detectives salvajes*', *MLN*, Vol. 125, nº 2 (2010), 406–432.

Sauri, Emilio, 'Autonomy after Autonomy, or The Novel Beyond Nation: Roberto Bolaño's *2666*', *Canadian Review of Comparative Literature*, 42 (2015), 396–409.

Schwartz, Jorge, 'Saer en la Universidad de Sao Paulo', *Avisora* [n.d.] <http://www.avisora.com/publicaciones/monosavizora/entrevista_saer.htm> [accessed 3 March 2008].

Shaffer, Elinor, 'Secular Apocalypse', in *Apocalypse Theory and the Ends of the World*, ed. Malcolm Bull (Oxford and Cambridge, MA: Blackwell, 1995), 137–154.

Shklovskly, Víktor, *Theory of Prose*, trans. Benjamin Sher (Elmwood Park, IL: Dalkey Archive Press, 1990).

Shuttleworth, Sally, '"So Childish and So Dreadfully Un-Childlike": Cultural Constructions of Idiocy in the Mid-Nineteenth Century', in *Crossing Boundaries: Thinking Through Literature* (Sheffield: Sheffield Academic Press, 2001), 17–44.

Smith, Anna, *Julia Kristeva: Readings of Exile and Estrangement* (Houndmills and London: Macmillan Press, 1996).

Sommer, Doris, *Foundational Fictions: The National Romances of Latin America* (Berkeley and Los Angeles: University of California Press, 1991).

Sontag, Susan, 'Questions of Travel', in *Where the Stress Falls* (London: Vintage, 2003), 274–284.

Spitta, Silvia, *Between Two Waters: Narratives of Transculturation in Latin America* (College Station: Texas A&M University Press, 2006).

Spurr, David, *The Rhetoric of Empire: Colonial Discourse in Journalism, Travel Writing, and Imperial Administration* (Durham, NC and London: Duke University Press, 1993).

Svendsen, Lars, *A Philosophy of Boredom* (London: Reaktion Books, 2005).

Todorov, Tzvetan, 'The Origin of Genres', *New Literary History*, Vol. 8, No. 1 (1976), 159–170.

Van Den Abbeele, George, *Travel as a Metaphor: From Montaigne to Rousseau* (Minneapolis: University of Minnesota Press, 1992).

Vezzetti, Hugo, 'La nave de los locos de Juan José Saer', *Punto de vista*, 59 (1997), 39–41.

Vila-Matas, Enrique, 'Bolaño en la distancia', in *Roberto Bolaño: la escritura como tauromaquia*, ed. Celina Manzoni (Buenos Aires: Corregidor, 2006), 97–104.

Vila-Matas, Enrique, 'Un plato fuerte de la China destruida', in *Bolaño salvaje*, eds Edmundo Paz Soldán and Gustavo Faverón Patriau (Barcelona: Candaya, 2008), 45–51.

Villalobos-Ruminott, Sergio, 'A Kind of Hell: Roberto Bolaño and the Return of World Literature', *Journal of Latin American Cultural Studies*, Vol. 18, 2–3 (2009), 193–205.

Viñas, David, *Indios, ejército y frontera* (Buenos Aires: Santiago Arcos Editor, 2003).

Villoro, Juan, 'El copiloto del Impala', in *Roberto Bolaño: la escritura como tauromaquia*, ed. Celina Manzoni (Buenos Aires: Corregidor, 2006), 77–80.

Volpi, Jorge, *El insomnio de Bolívar: cuatro consideraciones intempestivas sobre América Latina en el siglo XXI* (Buenos Aires: Debate, 2009).

Walker, Carlos, 'El tono del horror: *2666* de Roberto Bolaño', *Taller de letras*, 46 (2010), 99–112.

Weber, David J., and Jane M. Raush, *Where Cultures Meet: Frontiers in Latin American History* (Wilmington, DE: SR Books, 1994).

Wilson, Jason, 'Epic Elegy for a Lost Generation', *The Independent*, 24 August 2007 <http://www.independent.co.uk/arts-entertainment/books/reviews/the-savage-detectives-by-roberto-bolantildeo-trans-natasha-wimmerbrlast-evenings-on-earth-by-roberto-bolantildeo-trans-chris-andrews-462725.html> [accessed 10 October 2010].

Wilson, Jason, 'Travel literature', in Verity Smith, ed., *Encyclopedia of Latin American Literature* (Chicago: Fitzroy y Dearborn; Taylor and Francis, 2007), 803.

Wright, Melissa W., 'Feminine Villans, Masculine Heroes, and the Reproduction of Ciudad Juárez', *Social Text*, 69 (2001), 93–113.

Index

Iberian and Latin American Studies: The Arts, Literature and Identity

Edited by Francis Lough
Professor of Hispanic Studies, Department of Modern Languages, University of Birmingham

This series publishes titles from any area of Iberian and Latin American Studies that explore issues relating to questions of identity. The series accepts for publication scholarly monographs and collections of essays that aim to further our knowledge and understanding of the lives of individuals and communities who speak any of the languages of the Iberian Peninsula or Latin America. Ideas and concepts of identity can be explored at various levels, ranging from the individual to the national or international, and in different media. Proposals are welcome from researchers working in any cultural field, for example, the history of ideas, literature, performance, cinema, art and photography, and on a variety of issues, including nationhood, exile, memory, and gender. The series welcomes manuscripts in English or Spanish.

Those interested in contributing to the series are invited to write with the synopsis of a subject already in typescript or with a detailed project outline to either Professor Francis Lough, Department of Modern Languages, University of Birmingham, Edgbaston, Birmingham, B15 2TT, United Kingdom, f.lough@bham.ac.uk, or to Peter Lang Ltd, 52 St Giles, Oxford, OX1 3LU, United Kingdom, oxford@peterlang.com.

Vol. 1 Alison Ribeiro de Menezes and Catherine O'Leary (eds):
Legacies of War and Dictatorship in Contemporary Portugal and Spain
ISBN 978-3-03911-872-4

Vol. 2 Wendy-Jayne McMahon:
Dislocated Identities: Exile and the Self as (M)other in the Writing of Reinaldo Arenas
ISBN 978-3-0343-0223-4

Vol. 3 Helena Buffery (ed.):
Stages of Exile: Spanish Republican Exile Theatre and Performance
ISBN 978-3-0343-0263-0

Vol. 4 Erica Segre (ed.):
Ghosts of the Revolution in Mexican Literature and Visual Culture: Revisitations in Modern and Contemporary Creative Media
ISBN 978-3-0343-0702-4

Vol. 5 Daniela Omlor:
Jorge Semprún: Memory's Long Voyage
ISBN 978-3-0343-0821-2

Vol. 6 Maura Rossi:
La memoria transgeneracional: Presencia y persistencia de la guerra civil en la narrativa española contemporánea
ISBN 978-3-0343-2263-8

Vol. 7 Maria-José Blanco and Claire Williams (eds):
Feminine Singular: Women Growing Up through Life-Writing in the Luso-Hispanic World
ISBN 978-3-0343-0836-6

Vol. 8 Florencia Garramuño:
The Opaque Experience: Literature and Disenchantment
ISBN 978-1-78707-366-1

Vol. 10 Guadalupe Gerardi:
Writing Travel: The Work of Roberto Bolaño and Juan José Saer
ISBN 978-3-0343-2215-7

www.ingramcontent.com/pod-product-compliance
Ingram Content Group UK Ltd.
Pitfield, Milton Keynes, MK11 3LW, UK
UKHW020159290926
15471UKWH00005B/966

9 783034 322157